INTERNATIONAL SERIES OF MONOGRAPHS IN
LIBRARY AND INFORMATION SCIENCE
GENERAL EDITOR: G. CHANDLER

VOLUME 2

SOURCES OF INFORMATION ON
ATOMIC ENERGY

INTERNATIONAL SERIES OF MONOGRAPHS IN LIBRARY AND INFORMATION SCIENCE

Vol. 1. WHITE—*Bases of Modern Librarianship*

SOURCES OF INFORMATION ON ATOMIC ENERGY

BY

L. J. ANTHONY

Librarian,
U.K.A.E.A., Culham Laboratory,
Abingdon, Berks.

PERGAMON PRESS
OXFORD · LONDON · EDINBURGH · NEW YORK
PARIS · FRANKFURT

Pergamon Press Ltd., Headington Hill Hall, Oxford
4 & 5 Fitzroy Square, London W.1

Pergamon Press (Scotland) Ltd., 2 & 3 Teviot Place, Edinburgh 1

Pergamon Press Inc., 44–01 21st Street, Long Island City, New York 11101

Pergamon Press S.A.R.L., 24 rue des Écoles, Paris 5e

Pergamon Press GmbH., Kaiserstrasse 75, Frankfurt-am-Main

Copyright © 1966
Pergamon Press Ltd.

First edition 1966

Library of Congress Catalog Card No. 65–19990

AUTHOR'S NOTE

The Office of Technical Services of the U.S. Department of Commerce, Washington 25, U.S.A. is mentioned frequently in these pages. Since this text was prepared, however, a major reorganization has taken place in American Government Documentation Services and OTS has become the Clearing House for Federal Scientific and Technical Information, National Bureau of Standards, U.S. Department of Commerce, Springfield, Virginia, U.S.A.

CONTENTS

Introduction   ix

PART 1. ATOMIC ENERGY
Chapter 1. Atomic Energy   3

PART 2. NATIONAL SOURCES OF INFORMATION
Chapter 2. Information Sources in the British Commonwealth   13
    United Kingdom   13
    Canada   24
    Australia   26
    India   27
    Pakistan   29

Chapter 3. Information Sources in the United States   33
    Atomic energy research in the U.S.A.   33
    USAEC information services   35
    Other sources of information in the U.S.A.   41

Chapter 4. Information Sources in the Soviet Bloc   46
    Nuclear research in the U.S.S.R.   46
    Sources of Russian information   47
    Other Soviet bloc countries   51

Chapter 5. Information Sources in Other Countries   55
    France   55
    Italy   61
    Japan   65
    Federal Republic of Germany   69
    Sweden   73
    Belgium   76
    Netherlands   79
    Switzerland   83
    Norway   86
    Denmark   87

PART 3. INTERNATIONAL SOURCES OF INFORMATION
Chapter 6. International Organizations   97
    International Atomic Energy Agency   97
    European Atomic Energy Community   99
    European Nuclear Energy Agency   104
    European Organization for Nuclear Research   108
    European Atomic Forum   111

PART 4. PUBLISHED SOURCES OF INFORMATION

Chapter 7. Atomic Energy in General — 117
Bibliographies and abstracting services — 117
Directories and handbooks — 120
Encyclopaedias — 122
Dictionaries and glossaries — 123
Books — 123
Conference proceedings — 126
Translations — 128

Chapter 8. Atomic, Nuclear and High Energy Physics — 141
Atomic physics — 141
Nuclear physics — 142
Fundamental particles and high energy physics — 146
Nuclear data — 150

Chapter 9. Nuclear Power and Engineering — 159
Energy resources and nuclear power programmes — 159
Economic aspects of nuclear power — 160
Nuclear reactors — 161
Reactor theory — 163
Reactor control and instrumentation — 164
Reactor catalogues — 164
Nuclear propulsion — 165
Reactor safety — 165
Chemical processing and reactor materials — 166
Disposal of radioactive wastes — 169

Chapter 10. Ionizing Radiations and Radioisotopes — 177
General — 177
Separation and production of isotopes — 178
Applications of isotopes and radiation — 178
Directories and catalogues of isotopes — 181
Isotope laboratory techniques — 182
Large radiation sources — 182
Effects of radiation on materials — 183
Biological effects of radiation — 184
Protection against radiation — 187
Radiation counters and detectors — 190

Chapter 11. Controlled Nuclear Fusion and Plasma Physics — 203
Introduction — 203
Organization of fusion research — 204
Published information — 208

Subject Index — 219

Index to Organizations — 221

Author Index — 227

Index to Periodicals — 243

INTRODUCTION

Few fields of science and technology remain unaffected by the development of atomic energy and the need for a guide to its literature and to the organizations which originate atomic energy information has steadily increased.

The difficulties which face scientists, engineers, librarians and information officers in keeping track of the vast amount of scientific and technical information available are well known and will not be elaborated. These difficulties are particularly exacerbated in such subject fields as nuclear energy where the volume of "published" information, in the form of books and journal articles, is probably exceeded by that of "unpublished" information in the form of reports, internal memoranda, preprints and conference papers which never appear in published proceedings. Even when their existence is discovered these documents may not be readily available to people working outside the national nuclear energy projects, although the larger of these projects, such as those of the United Kingdom and the United States, have gone to some trouble to make their documents available as widely as possible by depositing complete collections in libraries open to the public or by putting them on sale through national agencies.

Although it is hoped that everyone seeking information in the nuclear energy field will find this guide useful, it has been written primarily with the needs of librarians and information officers in mind since they are often the first people to be approached when information is needed. It is mainly for this reason that an effort has been made in Chapter 1 to describe, in fairly simple terms, the various aspects of atomic energy and to show how they are related to each other and to other technologies. Some of these ideas are elaborated in Part 4, which deals with published literature, wherever it was felt that a brief explanation of the technical background would be helpful. Such interpolations have been made as short as possible and should not distract the scientific or technical reader who wishes to use the book as a straightforward guide to sources.

No guide of this kind can hope to be exhaustive. Nevertheless every effort has been made to include all important sources, weter they be organizational or documentary, and to describe these as accurately as possible within the limits of knowledge available up to August 1964. No doubt there are omissions and errors and the author would be glad to have these brought to his attention.

The opportunity is taken to thank the many friends and colleagues in both the United Kingdom Atomic Energy Authority and the United States Atomic Energy Commission who have volunteered constructive criticism of the material presented herein and particularly those who have so ably assisted in the preparation of the manuscript.

PART 1

ATOMIC ENERGY

CHAPTER 1

ATOMIC ENERGY

Energy Sources

The development of modern civilization is wholly dependent upon a plentiful supply of energy, which, until recently, has been obtained almost entirely from the fossil fuels, coal and oil, and from wind and water power. All these may be regarded as forms of stored solar energy as they exist by virtue of the effects of the sun's radiation.

Wind and water power, of course, are forms of mechanical energy whereas that derived from the fossil fuels is chemical in the sense that it is obtained as a result of chemical reactions which take place during burning. The most important of these reactions are the oxidation of carbon to form carbon dioxide and the oxidation of hydrogen to form water, both of which are exothermic, releasing energy in the form of heat. According to the Special Theory of Relativity this energy can be created only at the expense of matter and, in fact, when hydrogen and oxygen combine to form water with the evolution of heat, the mass of water formed is less than the combined masses of the reactants. In chemical reactions, however, this mass difference is so small that for all practical purposes it can be said that, if chemical energy only is involved, the mass of the reacting system remains constant. This is to be expected because in chemical reactions only the outer electrons of the atoms are involved and the nuclei, where most of the mass is concentrated, retain their identity.

The fossil fuels were formed millions of years ago as a result of climatic conditions which are unlikely to be repeated and the deposits are not inexhaustible. Various estimates of the extent of world reserves of coal and oil suggest that they will be used up within a few hundred years, assuming that the world demand for energy continues to rise at its present rate. Bearing in mind that coal and oil are valuable chemicals as well as fuels, there is clearly a pressing need for alternative energy sources in the near future and it is this need which has supplied the main impetus for the rapid development of research into the peaceful uses of atomic energy.

Nuclear energy is derived from interactions between the nuclei of atoms whereby mass is converted into energy according to Einstein's formula. In this case, however, the mass difference is of the same order of magnitude as the masses of the individual nuclei so that the amount of energy liberated per gram of matter is millions of times greater than that obtained from chemical reactions.

Nuclear Energy

The production of energy from the fusion of the nuclei of light elements was first demonstrated by Cockcroft and Walton in an experiment made in 1932 at the Cavendish Laboratory, Cambridge. They caused a stream of protons to bombard lithium atoms and succeeded in obtaining helium nuclei with energies greater than those of the incident protons. Calculations of the masses of the nuclei involved showed that the mass of the two helium nuclei produced per reaction was less than the combined mass of the proton and the lithium nucleus by an amount equivalent to the excess energies observed.

It is now believed that fusion reactions of a similar kind take place in the sun's interior and are in fact the source of the sun's seemingly unlimited energy. It has been calculated that in order to provide this energy the sun is losing mass at the rate of four million tons every second but the total mass of the sun is so great that it could continue to provide this stream of energy for something like 30,000,000,000 years. So far energy from nuclear fusion has been produced on earth only in the form of the hydrogen bomb. Experiments aimed at providing a controlled release of energy from the fusion of hydrogen isotopes have been going on since 1946 and are discussed in more detail in Chapter 11.

Energy is also produced by the disintegration of the nuclei of heavy elements, a process which has long been familiar as radioactive decay. Unfortunately this process is a random one, unaffected by heat, light or any other agent that man can control and therefore not suitable as a practical large-scale source of energy. In radioactive decay only a small fragment of the original nucleus is ejected and the reaction is not self-sustaining. There is, however, one kind of disintegration reaction which is both self-sustaining and controllable, namely the neutron-induced fission of uranium and it is this process which is the basis of existing nuclear science and technology.

Nuclear Fission

Natural uranium consists of a mixture of uranium isotopes of which only one, uranium-235, is fissile. This is a rather rare isotope which is normally present in the proportion of only one part in one hundred and forty. When bombarded by neutrons the U^{235} nucleus splits into two roughly equal parts which are nuclei of medium-weight elements, with the release of energy and several new neutrons which, in turn, can cause fission in other U^{235} nuclei thus setting up a chain reaction which is self-sustaining. The neutrons released in fission are extremely energetic and in a mass of natural uranium such high energy neutrons are more likely to be captured by atoms of the more abundant isotope U^{238}, than to cause fission in U^{235} atoms. It is found, however, that if the neutrons are slowed down the probability of a U^{235} nucleus capturing a neutron and undergoing fission is greater and under these conditions a chain reaction can be maintained in natural uranium with its small proportion of U^{235} atoms. This slowing down can be achieved by allowing the neutrons to strike the nuclei of some light element such as carbon, so that in each collision some energy is lost until eventually the neutrons are in thermal equilibrium with the atoms of the light element.

Elements, such as carbon, beryllium and hydrogen, which are used for this purpose are called moderators. Some light elements, such as boron, are not suitable for this purpose because they readily absorb neutrons and would stop the chain reaction. Such elements can be used to control the reaction by absorbing excess neutrons thus reducing the energy output to the level required.

Not all of the neutrons produced in fission reactions will be able to cause further fission. Some will be absorbed by U^{238} atoms to produce plutonium, another fissile material; some will be captured by U^{235} atoms without producing fission; some will escape altogether. The number of neutrons escaping will depend partly on the size of the lump of uranium in which the reaction is taking place: if this is too small more neutrons will escape than will cause fission and the reaction will die out. There is therefore a critical size which must be exceeded if the reaction is to be maintained.

Nuclear Reactors

The parameters which must be considered, therefore, when designing a nuclear reactor include the neutron speed, the degree of enrichment of the uranium fuel (i.e. the amount by which the proportion of U^{235} in the uranium is greater than normal), the shape and size of the lump of fuel used, the type of moderator and the purpose for which the reactor is required.

Neutron speed has been used as the primary characteristic for classifying fission reactors, the main divisions being fast reactors, intermediate reactors and thermal reactors. The vast majority of reactors use a moderator and therefore fall into the last category. Thermal reactors are classified by moderator and by coolant although in some cases these may be one and the same material.

The energy made available in the fission reaction appears as kinetic energy of the fission fragments which, in their passage through the reactor core, generate heat. This heat is carried away by the coolant and can be used to generate electricity by conventional methods. Power reactors are designed so that the maximum amount of useful heat is obtained at as high a temperature as possible. Indeed much of the recent research on power reactors has been concerned with the design of materials and core systems which will provide heat at higher temperatures and thus provide electrical power more economically.

By varying the basic parameters a wide range of reactors can be designed, each suitable for a particular purpose. Reactors are the source of a plentiful supply of energetic neutrons and can therefore be used as a research tool by nuclear physicists or for testing the effects of radiation on materials. Other purposes for which reactors have been built include the production of radio-isotopes, breeding fissile materials, preserving food and drugs by irradiation, providing heat for chemical processing and nuclear propulsion.

Radioisotopes

The number of protons in the nuclei of an element determines the number of orbiting electrons in the atoms of that element and hence its chemical properties. For any particular element, therefore, the number of protons in the nucleus is fixed. The number of neutrons, however, may vary within certain limits, depending on the elements being considered. Atoms of an element which contain the same number of protons but differing numbers of neutrons in their nuclei are called isotopes of the element: they are indistinguishable chemically but have different masses.

Many elements exist in nature as a mixture of two or more isotopes and in most of these cases the isotopes are stable. If, however, a quantity of an element is subjected to the intense neutron flux in the core of a reactor, nuclear reactions will take place and isotopes of other elements will be formed. In many cases these isotopes are artificial in the sense that they do not exist in nature and most of them are highly radioactive. Such isotopes are sources of intense radiation which, by suitable shielding, can be controlled and many uses have been found for these substances in medicine, biology, agriculture and industry.

In general such uses fall into three categories. The isotope may be used as a convenient source of radiation as, for example, in the treatment of cancers. Isotope sources of this kind have been used for irradiating food in order to preserve it; for sterilizing medical utensils; as thickness gauges in industrial processing; and for research into the effects of radiation on materials and tissues.

Because a radioisotope is constantly emitting radiation which can be detected by suitable instruments, it can readily be used as a tracer or indicator. For example by mixing a small quantity of radiophosphorus with phosphate fertilizers it is possible to discover how and when plants absorb phosphorus and what happens to it in the plant. In this way it has been shown that certain plants, such as sugar beet, need phosphorus only at the seedling stage, whereas other plants like the potato need it throughout the growing period of the plant. By labelling elements with radioactive isotopes a great deal has been discovered about the uptake of nutrients in plants and animals and tracer techniques are now an essential part of biological and medical research. The same principles can be used to locate leaks in water pipes, measure the wear of metals in piston rings, trace the movement of mud and gravel in estuaries and other waterways, test the effectiveness of a detergent and determine the age of fossils.

Radioisotopes can also be used as power generators. The radiations from a small amount of radioactive material, sealed in a container, are absorbed by the container, thus generating heat. By thermoelectric or thermionic means the heat can be converted directly into electricity and the device used as a small scale power source. Such a device, about the size of a grapefruit, and capable of delivering 11,000 W-hr of electricity over a period of about 280 days was first demonstrated in America in January, 1959. Such compact devices can supply power for prolonged periods without attention and are being developed for use in rockets and satellites and as a means of providing power for unmanned weather stations and similar facilities in remote regions.

The use being made of isotopes increases every day and radioisotope production is now a commercial undertaking in Britain and America which, at present, are the world's main suppliers.

Ionizing Radiation

Radioisotopes, like the fragments which are produced in a nuclear reactor as a result of uranium fission, are radioactive, emitting ionizing radiation in the form of alpha particles, beta particles or gamma rays. This radiation, together with the neutrons which are also produced in the reactor, has a destructive effect on living tissue so that the core of the reactor must be shielded in order to protect the people working with it. Similar protection must be afforded to people using radioisotopes for medical or industrial purposes. Protection against ionizing radiation is an important aspect of nuclear science and technology and figures largely in the atomic energy literature. Research is aimed at finding the most effective shielding materials; devising means for the remote handling of radioactive substances so that they can be manipulated, tested and analysed; determining the effects of radiation on living cells and seeking ways of counteracting these effects. In general the effects are proportional to the magnitude of the dose received and much effort at national and international level has been devoted to designing safety regulations and codes of practice which will ensure, under all circumstances, that permissible doses are not exceeded.

The rate at which radioactive substances emit radiation varies considerably and for any particular isotope is independent of the environment. Each isotope has a characteristic decay period known as the half-life which is defined as the time taken for the radioactivity of any given quantity of the isotope to decay to half its initial value. This implies that, given a long enough period, the activity of any quantity of an element will be reduced to a very small amount but can never, theoretically at any rate, become zero. The half-lives of the radioisotopes vary from fractions of a second to millions of years. Polonium-213, for example, has a half-life of approximately four millionths of a second, whereas that of thorium-232 is nearly fourteen thousand million years. The characteristic decay period, or half-life, identifies an isotope as clearly as a fingerprint identifies a person, and this property is the basis of the very powerful chemical analysis technique known as activation analysis which can detect elements which are present in quantities as low as one thousand millionth of a gram.

This, of course, is not the only area in which chemistry and nuclear physics meet. Ionizing radiations have profound chemical and structural effects on materials, some inimical, some useful. For example, ionizing radiation can be used to polymerize certain organic compounds and thus create new plastic materials. The study of the chemical effects of radiations is the field of radiation chemistry, a term which is sometimes confused with radiochemistry which is the study of the chemical behaviour of radioactive compounds, particularly the production of such compounds by chemically processing irradiated or naturally radioactive materials.

Raw Materials and Processing

Chemical processing is necessary to prepare the fuel before putting it into the reactor, and to extract the fission products after the reaction has taken place.

The commonest uranium containing ores are pitchblende, from which the Curies first isolated radium, and carnotite which was once used by the North American Indians as war paint. Uranium is usually extracted from the ores in the form of the yellow oxide which is then purified and treated to produce uranium metal. Only small quantities of metal are obtained from many tons of ore and because of the high degree of purity required the process is fairly costly. All uranium compounds are radioactive and most are toxic so that stringent safety precautions are necessary and a good deal of research has been undertaken in order to devise methods for extracting high purity uranium as safely and as economically as possible.

Natural uranium is not suitable for weapons purposes nor as a fuel for many types of reactor so that means have to be found for extracting the fissile isotope U^{235} in a pure state. Chemical extraction is not possible since there is no difference chemically between U^{238} and U^{235} and the only characteristic which can be used to identify them is the very slight difference in mass.

Ingenious methods of separating the isotopes, using this property, have been devised, the most practical large-scale method being the gaseous diffusion process. Natural uranium in the form of uranium hexafluoride, which is a gas at temperatures slightly above normal, is made to diffuse through a series of membranes, the separation process depending on the fact that, at any given temperature, the molecules of a lighter gas move faster, on average, than those of a heavier gas, and hence pass through the pores in a membrane more quickly. In practice about four thousand stages are needed to obtain a high degree of enrichment, and the number of pumps and compressors required to keep the gases circulating is so great that one single factory of this kind can consume all the electricity generated by a large, modern power station. At the present moment only the United Kingdom, the Unites States and the Soviet Union have gaseous diffusion plants, although a similar plant is being designed in France.

In a reactor neutrons are captured by the U^{238} atoms in the fuel to produce U^{239} which decays rapidly by emission of beta particles to form neptunium-239. This in turn decays, again by beta emission, to form plutonium-239 which is relatively long-lived and fissile and can therefore be used for weapons and as a reactor fuel. It is therefore necessary to design processing plant in which the plutonium can be extracted from the fuel elements which have completed their useful life in the reactor. The extraction can be done chemically and would be fairly straightforward were it not for the fact that the spent fuel is highly radioactive and most of the manipulation must be carried out by remote control in heavily shielded chambers. Plutonium itself is, of course, radioactive and highly toxic so that elaborate precautions are necessary to safeguard those working with it.

After the plutonium has been extracted there remains the problem of dealing with the highly radioactive fission product wastes. Some fission products have useful applications but the bulk of the waste must be disposed of without creating a radiation hazard. Short-lived fission products can be stored in tanks until the radioactivity has decayed to the point where it is safe to pump the residue into rivers or the sea. Longer-lived isotopes can be converted into ceramic materials from which the radioactivity cannot be washed out and these solid materials can be buried or dumped in the deeper parts of the ocean. The problem of waste disposal is by no means solved and is one of the reasons for hoping that the comparatively " clean " fusion process will replace fission energy in the not too distant future.

Nuclear and High Energy Physics

The realization in 1938, of the nature of the fission process was a direct consequence of a chain of events dating back to Becquerel's discovery of radioactivity in 1896 and covering a period during which the groundwork of nuclear physics was laid. Since 1938 the task of developing satisfactory theoretical models to account for the phenomena associated with nuclear energy has enormously stimulated research in nuclear physics, while at the same time the development of nuclear energy has provided new tools to help the nuclear physicist in his work.

The fundamental problem is to obtain an understanding of the structure of the nucleus and of the forces that operate within it, particularly the very strong short-range forces which hold the neutrons and protons together against the electrostatic repulsive forces tending to disrupt the nucleus. The short-range forces are essentially different from the more familiar forces such as those of gravitation and electrostatics and are not yet understood.

A powerful method for obtaining information about the nucleus is to bombard chosen nuclei with high energy particles and study the resulting reactions. The energies of the incident particles must be extremely large if they are to penetrate the potential barriers of the nucleus and hence the need for larger and more powerful particle accelerators in which protons, electrons and other charged particles can be accelerated to speeds approaching that of light in order to obtain the energies required.

Although high energy accelerators are indispensable tools for the nuclear physicist, they are large, complex and very expensive and therefore beyond the resources of most universities and research organizations. For the very largest the cost is so high that they can be provided only on a national or even international basis. A good example of the latter is the proton-synchrotron built at Geneva by the European Organization for Nuclear Research (CERN) which was established in 1954 by twelve European countries, each of which makes an annual financial contribution.

The beams of high energy particles produced in accelerators can be used to irradiate quantities of an element or compound and thus create radioisotopes, and, in fact, this is the only way in which some isotopes can be made. The main purpose of the accelerators, however, is as a tool for research in the field of high energy physics, where they are the main hope of obtaining answers to the fundamental problem of the nature of matter.

PART 2

NATIONAL SOURCES OF INFORMATION

The development, organization and activities of the major national atomic energy projects and other national organizations concerned with atomic energy are described in the next four chapters. Chapter 2 covers the United Kingdom and those Commonwealth countries which have well-developed atomic energy programmes. The main sources of information in the United States are described in Chapter 3 and Chapter 4 gives some account of the atomic energy organization in the Soviet Union and some of the smaller countries in the Soviet Bloc. Of the remaining countries only a comparatively small number have reached the stage where their national organizations can be regarded as important sources of information and these are described in Chapter 5.

CHAPTER 2

INFORMATION SOURCES IN THE BRITISH COMMONWEALTH

UNITED KINGDOM

The development of the present atomic energy organization in the United Kingdom began in 1945 when the British scientists, who had been working on the atomic bomb project in the United States and Canada, returned to Britain. At that time atomic energy research was the responsibility of the Department of Scientific and Industrial Research but after the passing, in America, of the MacMahon Act in 1946, which for the time being closed the door on Anglo-American co-operation in this field, it became necessary for Britain to develop an independent weapons programme and under the Atomic Energy Act of 1946 responsibility was transferred to the Ministry of Supply. Hence the first atomic energy establishments to be set up were entirely Government owned and controlled and for the first five years were fully engaged on defence commitments.

As the emphasis shifted from the military to the peaceful uses of atomic energy, industry began to be drawn in, at first by undertaking research and development contracts for the Ministry, and later, when the nuclear power programme got under way, by forming consortia fully equipped to design and build nuclear power stations and research reactors. Over a thousand British firms are now directly involved either in the manufacture of nuclear equipment or in the development of nuclear technology, and atomic energy has become a major industry in which public authorities, government departments, private firms, universities and many other organizations participate.

Atomic energy in Britain is not yet twenty years old, and a full account of its history, organization and development is yet to be written. A broad picture can be obtained from a booklet called *Nuclear Energy in Britain* [1] issued by the Central Office of Information to meet the needs of the Overseas Information Service. In ninety pages the booklet reviews the development of atomic energy in Britain, the organizations responsible, the evolution of the nuclear power programme and the part played by industry, current nuclear research and Britain's part in international collaboration.

The United Kingdom Atomic Energy Authority

The official body for atomic energy research and development in the United Kingdom is the U.K. Atomic Energy Authoritiy, a public corporation established by the *Atomic Energy Authority Act, 1954* to take over the responsibility for atomic energy hitherto borne by the Ministry of Supply. The Authority are responsible to the Minister of Technology who appoints the members but the structure of the Authority is more akin to that of a big industrial organization than to that of a government department. The Authority have very wide powers in the atomic energy field, their main functions being to carry out research into the peaceful uses of atomic energy; to assist in the fulfilment of the United Kingdom nuclear power programme and to select and develop reactor systems for this purpose; to produce and market radioisotopes and nuclear fuels; and to develop weapons in accordance with the requirements of the Ministry of Defence.

To this end the Authority are organized into five Groups, each of which enjoys a considerable measure of autonomy in its day-to-day operations. These are the Research Group based on Harwell, Berkshire; the Reactor, Production and Engineering Groups based on Risley, Lancashire; and the Weapons Group based on Aldermaston, Berkshire. In addition to the five Groups and independent of them is the Authority Health and Safety Branch with divisions in London, Harwell and Risley. The Authority also maintain a London Office to provide central administrative and secretarial services. A more detailed account of the history and organization of the Authority will be found in *The United Kingdom Atomic Energy Authority: its history and organisation* [2] which can be obtained from the London Office.

Each of the Groups of the Authority has a specific function within the general programme of the Authority. In the Research Group the emphasis is on fundamental studies in the nuclear sciences, early development work on power generation, testing and evaluation of materials and the production and use of radioisotopes. Research and development in the field of fission reactors is the main function of the Reactor Group, probably the largest single unit in the world specializing in this field. The Weapons Group undertakes fundamental research in nuclear physics as well as being responsible for the design and development of nuclear weapons, while the production of uranium und plutonium and the manufacture of fuel elements for the Authority's reactors and for the civil nuclear power stations, is the main concern of the Production Group. The Engineering Group provides a service to the other groups in the design and construction of plant and buildings and is also concerned with the design and inspection of fuel elements.

Within this broad outline each of the Authority's establishments carries out a research and development programme the main features of which are summarized below.

Research Group

Atomic Energy Research Establishment, Harwell, Berkshire. Founded in 1946, this is the oldest of the Authority's laboratories and probably the best known, nationally and internationally. It is one of the largest laboratories in the United Kingdom with a staff of over 6000 and its work includes basic

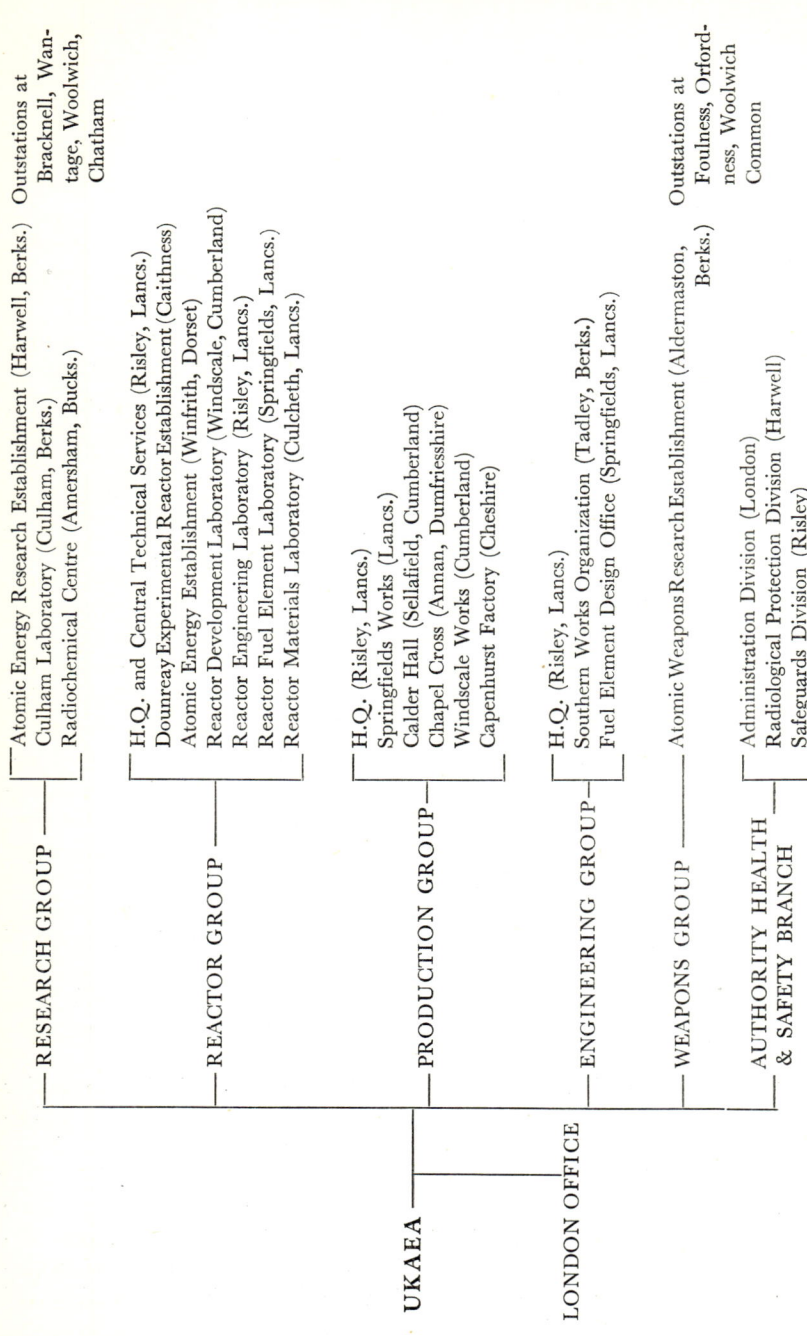# ORGANIZATION OF THE UKAEA

UKAEA
- LONDON OFFICE
- RESEARCH GROUP
  - Atomic Energy Research Establishment (Harwell, Berks.) — Outstations at Bracknell, Wantage, Woolwich, Chatham
  - Culham Laboratory (Culham, Berks.)
  - Radiochemical Centre (Amersham, Bucks.)
- REACTOR GROUP
  - H.Q. and Central Technical Services (Risley, Lancs.)
  - Dounreay Experimental Reactor Establishment (Caithness)
  - Atomic Energy Establishment (Winfrith, Dorset)
  - Reactor Development Laboratory (Windscale, Cumberland)
  - Reactor Engineering Laboratory (Risley, Lancs.)
  - Reactor Fuel Element Laboratory (Springfields, Lancs.)
  - Reactor Materials Laboratory (Culcheth, Lancs.)
- PRODUCTION GROUP
  - H.Q. (Risley, Lancs.)
  - Springfields Works (Lancs.)
  - Calder Hall (Sellafield, Cumberland)
  - Chapel Cross (Annan, Dumfriesshire)
  - Windscale Works (Cumberland)
  - Capenhurst Factory (Cheshire)
- ENGINEERING GROUP
  - H.Q. (Risley, Lancs.)
  - Southern Works Organization (Tadley, Berks.)
  - Fuel Element Design Office (Springfields, Lancs.)
- WEAPONS GROUP
  - Atomic Weapons Research Establishment (Aldermaston, Berks.) — Outstations at Foulness, Orfordness, Woolwich Common
- AUTHORITY HEALTH & SAFETY BRANCH
  - Administration Division (London)
  - Radiological Protection Division (Harwell)
  - Safeguards Division (Risley)

research in physics and chemistry, particularly the obtaining of accurate and reliable nuclear data, studies of the chemistry and metallurgy of materials, reactor experiments and the development of supporting technologies, and research into the health and safety aspects of atomic energy. Harwell is also the site of the Reactor School which, since 1954, has provided engineers and physicists from all over the world with basic training in the theory and operation of fission reactors.

The main research facilities at AERE are described in the *Harwell Handbook* [3] copies of which can be obtained from the Public Relations Section at Harwell.

Culham Laboratory, Culham, Abingdon, Berkshire. Research into plasma physics and controlled fusion, previously carried on at Harwell and Aldermaston, is now concentrated in the Culham Laboratory, the youngest of the Authority's establishments. The transfer of experimental work to Culham started in 1961 and will be completed in 1965.

Radiochemical Centre, Amersham, Buckingham. The Radiochemical Centre is responsible for the preparation and marketing of a wide range of radioisotopes, labelled compounds and radiation sources, which are distributed to more than 70 countries.

Wantage Research Laboratory, Wantage, Berkshire. This is an outstation of the Atomic Energy Research Establishment and houses the Isotope Research Division which undertakes research into, and development of, industrial applications of radioactive materials. The Isotope Branch of the Division runs an advisory service for industry on the use of radioisotopes and undertakes experimental work for customers either in its own laboratories or at the customer's own establishment. The Radiation Branch is concerned mainly with the use of strong radiation sources in such fields as sterilization and plants genetics.

Wantage also houses the Isotope School which runs a series of courses aimed at giving basic instruction in the measurement, handling and use of radioisotopes for those who wish to use them as research tools.

Reactor Group

Atomic Energy Establishment, Winfrith, Dorset. This is the major Authority establishment for research into experimental reactor systems. Formerly part of the Research Group it was transferred to the Reactor Group in 1961. Work includes theoretical and experimental aspects of neutron physics and reactor physics, heat transfer in reactor coolants, and reactor instrumentation and control. Winfrith is also the site of the high temperature, gas-cooled reactor *Dragon,* a project sponsored by the European Nuclear Energy Agency of the Organization for Economic Co-operation and Development in which Britain is co-operating.

Dounreay Experimental Reactor Establishment, Caithness. This is the site of the Authority's Fast Breeder Experimental Reactor designed to investigate the possibilities of fast reactors as power producers and to develop fast reactor fuels. The Establishment also has a materials testing reactor and provides a comprehensive irradiation service for the other establishments.

Reactor Development Laboratories, Windscale, Cumberland. The Laboratory's main function is development of the Advanced Gas Cooled Reactor and the testing and development of AGR fuel elements.

Reactor Materials Laboratories, Culcheth, Warrington, Lancashire. Carries out research into physical and chemical properties of reactor materials.

Reactor Engineering Laboratories, Risley, Lancashire. Engineering development of reactor components and irradiation test equipment is the main purpose of the laboratory which also does experimental work in connection with reactors for marine propulsion.

Reactor Fuel Element Laboratories, Springfields, Lancashire. Engaged in the development of fuels and fuel elements for thermal reactors.

Production Group

Springfields Works, Lancashire. Uranium ores and concentrates are processed to obtain uranium metal at this factory. From the metal so obtained fuel elements are manufactured for use in British and foreign reactors.

Calder Hall, Sellafield, Cumberland and Chapel Cross, Dumfriesshire. Both of these are sites of production reactors which, as well as producing plutonium, supply electricity to the national grid. Calder Hall is the prototype of the nuclear power stations now being commissioned by the Central Electricity Generating Board, and the Calder Operation School, opened in 1957, has trained most of the operating and commissioning staffs for the civil power stations in Britain as well as for the British designed stations in Japan and Italy.

Windscale Works, Cumberland. A factory where irradiated fuels are processed to separate the plutonium, uranium and fission products.

Capenhurst Works, Cheshire. This is the site of the gaseous diffusion plant which supplies the Authority with enriched uranium.

Weapons Group

Atomic Weapons Research Establishment, Aldermaston, Berkshire. Apart from research and development on explosive nuclear assemblies, the establishment undertakes fundamental research studies and development work in support of the civil nuclear power programme. Field trials in aid of weapons development are carried out at the outstations at Foulness, Essex and Orfordness, Suffolk.

Authority Health and Safety Branch

The Branch advises the Authority on their health and safety policy, carries out checks on reactors, plant and laboratories and provides a focal point from which the Authority's external relations in the health and safety field are conducted. The Safeguards Division at Risley is concerned primarily with the evaluation of the safety of nuclear reactors and processing plant while the Radiological Protection Division at Harwell studies the radiological and other hazards affecting personnel.

The research carried out by the Authority is extensive and varied and there is no single publication that adequately describes it. An account of the work of the Atomic Energy Research Establishment during its first eight years will be found in two books by Kenneth Jay [4, 5] both of which are sponsored by the Authority. Jay has also described the work of the Production Group [6] and the building of Britain's first nuclear power station at Calder Hall [7]. A more recent picture of the work of the Authority can be obtained from the annual reports [8] of the Authority to the Minister for Science, which are published as House of Commons papers, and from *Atom* [9] the monthly information bulletin of the Authority.

UKAEA Information Services

The administration of the Authority is largely decentralized so that there is no central control of the Authority's libraries and information services other than that exercised by the UKAEA Library and Information Services Coordinating Committee which co-ordinates policy. In general scientific and technical information services are provided by the Group libraries whereas information of a more general kind is provided by the Public Relations Branch in the London Office. The AERE Library is the main information centre in the Research Group and the Risley Library plays a similar role in the Reactor, Production and Engineering Groups. More specialized information services are provided by the libraries at Culham (plasma physics), and at Wantage and Amersham (isotopes). There is very close co-operation between the libraries and an enquiry directed to any one of them will be routed to the appropriate centre.

What is Available

It is the policy of the Authority to make known to industry and to the scientific world the results of their research and development work and this is done in a number of ways. Wherever possible information is made available in articles submitted to scientific and technical periodicals or in papers presented at scientific conferences. Approximately four hundred papers are published each year in this way. However, since much of the work of the Authority is highly specialized and because there are often delays in getting papers published in the open literature, a good deal of information is issued in the form of technical reports and similar documents. There are, of course, other reasons why it is not always possible to publish research results. A proportion of the information generated by the Authority is either commercially valuable or has defence implications and for either reason cannot be made generally available. Much of this information is put into reports, the distribution of which must be restricted or limited in some way, and this is true not only in the United Kingdom but in every country in which an official atomic energy project operates. Information which can be published without restriction or limitation is generally described as "unclassified" and it is with information of this kind only that this book is concerned.

Each Group of the Authority is responsible for its own publication programme and issues reports in numbered series which designate the originating establishment or group. As a result of changes which have taken place from time to time in the internal structure of the Authority, corresponding changes have had to be made in the numbering series used by the various groups and establishments and to the uninitiated the picture may sometimes seem a little confused. For this reason a full description of the report numbering series used since the early days of the atomic energy project has been issued under the title *Guide to UKAEA Documents* [10] which is available from the London Office and covers both current and obsolete report numbering series.

Since the beginning of the project the Authority libraries have used the Universal Decimal Classification for the subject classification of books, reports, pamphlets and other documents, and have played a large part in developing up-to-date schedules for the sections of the classification relating to nuclear physics and nuclear engineering. All reports issued by the Authority carry appropriate U.D.C. numbers and in order to obtain uniformity in the classification of the reports issued by the various establishments a Code of Practice was developed in 1959 which covers not only nuclear physics and engineering but all the sections of the U.D.C. which are of interest to the Authority. This Code of Practice is the basis of a special subject edition of the U.D.C. [11] issued by the International Federation for Documentation and jointly sponsored by the UKAEA and the International Atomic Energy Agency.

A number of booklets dealing with the work of the Authority generally or with specific aspects of it are issued from time to time, together with a series of *Information Papers* which give the full text of speeches, lectures and selected articles by Authority staff and other miscellaneous information. These are generally listed in *Atom* [9] at the time of publication and full details of all these can be obtained from the London Office. The Authority also produce films which can be borrowed by responsible organizations either from the Film Library of the Public Relations Branch in London or from U.K. Information Offices overseas. A catalogue [12] of these is available from the London Office which also maintains a comprehensive library of photographs and lantern slides.

Up-to-date information on recent developments in the Authority can be obtained from the press releases issued by the London Office. These are in two series [13, 14] covering technical developments, administrative changes and policy statements and are given wide distribution; they are also reproduced in the monthly issues of *Atom*. [9]

Dissemination of Information

All unclassified UKAEA publications are listed in the monthly *UKAEA List of publications available to the public* [15] which is published by the AERE Library on behalf of all the groups, and is sent regularly, without charge, to those who ask for it. The list is cumulated annually and the cumulations are sold by Her Majesty's Stationery Office.

The *UKAEA List* provides a complete record of publicly available reports and papers issued since December 1955. Before that date a considerable number of reports and articles had been written and published, most of them by the staff of the Atomic Energy Research Establishment and these are listed in three bibliographies [16, 17, 18] produced by AERE and sold by H.M.S.O. Apart from these general lists the Group Libraries issue bibliographies and reading lists on particular aspects of atomic energy and a complete list of these will be found in the *Guide to UKAEA documents*. [10] They are also listed in the United States Atomic Energy Commission publication *Bibliographies of interest to the atomic energy program* (TID–3043) (see Chapter 3, ref. 12), and its supplements. New bibliographies and revisions of old ones are listed, as they appear, in the *UKAEA List* [15] and notified to the USAEC so that they also appear in *Bibliographies of atomic energy literature issued or in progress* (see Chapter 3, ref. 13).

Harwell and Risley publish a weekly *Information Bulletin* and Culham Laboratory a *Library Bulletin* containing selected references to the world periodical literature in their fields of interest, and although these are intended mainly as a means of disseminating information internally, the lists are available to industrial and other organizations who would find them useful.

Copies of each unclassified report or bibliography are sent by airmail to the USAEC so that entries may appear in *Nuclear Science Abstracts* as early as possible. With the exception of those issued by the Atomic Weapons Research Establishment, reports are also sent to *Chemical Abstracts* and, selectively according to the subject matter, to *Analytical Abstracts, Metallurgical Abstracts, Science Abstracts,* and the *Journal of Applied Chemistry*.

The majority of the unclassified reports issued by the Authority are sold by H.M.S.O. and are therefore listed in the daily, monthly and annual lists of Government publications. The Stationery Office have also issued a sectional list [19] of atomic energy documents sold by them and this covers not only the publications of the Authority but those of a number of British government departments and international organizations for whose publications H.M.S.O. acts as a sales agency. For those who wish to purchase, in the form of a standing order, all Authority reports in one or more subject fields, a subscription service is operated by the AERE Library. The service includes all documents normally sold through H.M.S.O. as well as a small number of reports which, for one reason or another, are not put on sale, and subscriptions may be placed for documents in all or any of the twenty-two categories shown in Table 1.

TABLE 1. LIST OF CATEGORIES OF REPORTS AVAILABLE ON SUBSCRIPTION.

(These categories correspond to the broad subject headings under which reports are listed in the cumulations of the *UKAEA List of publications available to the public*)

1. General
2. Biology, medicine, agriculture
3. Chemistry—general
4. Chemistry—radio-, radiation and nuclear
5. Chemistry—analytical
6. Chemical engineering
7. Engineering
8. Instrumentation
9. Geology, mineralogy
10. Health and Safety
11. Industrial applications of isotopes and radiation
12. Isotope separation
13. Mathematics and computers
14. Metallurgy and materials
15. Radiation effects
16. Physics—general and solid state
17. Physics—nuclear and theoretical
18. Plasma physics and controlled fusion
19. Particle accelerators
20. Reactor physics
21. Reactor technology
22. Waste disposal

All unclassified reports issued between 1947 and 1956 including those sold by H.M.S.O. can be obtained in micro form from Micro Methods Ltd. of East Ardsley, Wakefield, Yorks, who also sell microcopies of unclassified reports issued since January 1957, except those sold by H.M.S.O. Further details can be obtained from Micro Methods' catalogue. [20]

UKAEA unclassified reports are deposited in eighteen of the larger public libraries in the United Kingdom from where they may be borrowed or photocopies obtained at the usual charges. Most of these depository libraries also receive reports regularly from the U.S.A., Canada, France, Italy, Norway, Denmark and other countries. A complete list of these libraries with a summary of their holdings of atomic energy reports will be found in the *Guide to UKAEA documents* [10] which also contains a list of the foreign atomic energy organizations, in forty countries, to which Authority documents are sent regularly.

In addition to the depository libraries, copies of all unclassified reports are sent to the Public Information Centre in the Authority's London Office where they may be referred to. The Centre, which includes an Isotope Information Bureau, is always ready to advise on the availability of documents and to put enquirers in touch with the appropriate sources.

Other Sources of Information in Britain

Science Research Council

The Science Research Council of the Department of Education and Science is now responsible for the nuclear physics research previously undertaken by the National Institute for Research in Nuclear Science, a body set up in 1957 to furnish facilities and equipment for carrying out research in nuclear physics which could be used by scientists from universities and other organizations, but which would be too costly or complex for individual universities to provide. The Institute's first laboratory, built on a site adjacent to the Atomic Energy Research Establishment at Harwell, was the Rutherford High Energy Laboratory, the principal features of which are a 7000 MeV proton synchrotron known as *Nimrod* which came into operation in 1963 and a 50 MeV proton linear accelerator which has been in use since 1959.

Information derived from experimental work at the Laboratory is normally published in scientific journals but the Laboratory does issue some unclassified reports, mainly concerned with aspects of the design and operation of accelerators, which are sold through the Stationery Office and deposited in the Science Museum Library, the National Lending Library for Science and Technology, and the public libraries in Birmingham and Leeds.

A second laboratory, to be called the Daresbury Nuclear Physics Laboratory, is now being built at Daresbury in Cheshire. The Laboratory will house a 4000 MeV synchrotron and is expected to be completed in 1966.

The most fruitful source of information about the Institute's operations are the annual reports [21] which contain detailed descriptions of the accelerators and of the research facilities available.

Central Electricity Generating Board

Research in the Central Electricity Generating Board is the responsibility of the Research and Development Department which operates three laboratories as part of the Headquarters research organization. These are the Berkeley Nuclear Laboratories situated on the site of the Berkeley nuclear power station in Gloucestershire, the Central Electricity Research Laboratories at Leatherhead in Surrey and the Marchwood Engineering Laboratories near Southampton. Of these Berkeley has been established to carry out the greater part of the Board's research and development work arising from the nuclear power programme. Research is concerned solely with the reactor aspects of nuclear power stations and covers materials research, reactor operation and safety, and fundamental studies aimed at providing information to aid in the design of future power stations. The work of the Laboratories is described in more detail in a CEGB booklet [22] issued in 1962 which also contains a list of papers published by the Laboratory staff.

The Board publishes an annual report [23] which contains a good deal of information on its research programme and a weekly *Digest* [24] prepared by the CEGB Information Services mainly for internal consumption, which contains abstracts of journal articles, recent additions to the Headquarters Library, and recent translations made by the Information Services.

British Nuclear Energy Society

A British Nuclear Energy Conference was formed in 1955 by the Institutions of Civil, Mechanical, Electrical and Chemical Engineers and the Physical Society, all of which had considerable interests in atomic energy. In 1962 the Conference was succeeded by the British Nuclear Energy Society, membership of which is open to all members of the constituent organizations and to others working in the nuclear field. The Society arranges meetings, symposia and conferences on a national and international basis and publishes a quarterly journal (see Appendix to Chapter 9), which contains papers given at meetings of the Society and other original contributions. The constituent bodies are now eleven in number, the five mentioned above together with the Iron and Steel Institute, the Institute of Metals, the Institute of Fuel, the Joint Panel on Nuclear Marine Propulsion, the British Institution of Radio Engineers and the Royal Institute of Chemistry.

Nuclear Energy Trade Associations' Conference

The Conference is a grouping of trade associations established to provide a permanent means of liaison between the many firms working in the atomic energy field. The membership comprises the British Chemical Plant Manufacturers' Association, the British Electrical and Allied Manufacturers' Association, the Scientific Instrument Manufacturers' Association of Great Britain, the Society of British Aircraft Constructors Ltd., the Water-tube Boilermakers' Association and the British Engineers' Association, the latter providing the secretariat at 32 Victoria St., London, S.W.1. The Conference is able to provide advice and information about appropriate firms and sources of supply for nuclear equipment.

Industrial Organizations

A number of industrial firms or groupings of firms have evolved, specializing in one aspect or another of atomic energy. The development of the nuclear power programme led to the formation of a number of Consortia of engineering firms, each of which has at its disposal all the resources needed to design and construct complete power stations. Between them the three Consortia have built, or are building, twelve nuclear power stations in Britain and overseas, and the individual companies which form the Consortia carry out extensive research and development programmes. In addition a number of firms specialize in the design and manufacture of research reactors for various purposes and in the manufacture of nuclear instruments and other ancillary equipment.

A good deal of information on the individual firms and their services will be found in *The Nuclear Energy Industry of the United Kingdom* [25] prepared by the UKAEA as a guide to overseas buyers.

British Nuclear Forum

Until very recently there has been no organization in Britain which could serve as a focal point for all nuclear interests in the country. Such an organization, called Atomic Industrial Forum, has existed in the United States since 1953 and has proved extremely useful as a meeting point at which industry, official organizations, societies and other bodies could discuss their common interests and co-ordinate their approach. This organization is described in Chapter 3.

The British Nuclear Forum was formed in December, 1963 with three classes of membership. These are (a) ordinary membership comprising industrial firms which will form the main body of the Forum; (b) association membership to cater for trade associations and professional bodies; (c) individual membership; and (d) associate membership to provide for Government organizations such as the UKAEA. The secretariat of the Forum is at present housed in the Federation of British Industries Offices at 21 Tothill Street, London S.W.1.

One of the objects of the Forum is to act as a channel for information, enquiries and expressions of view; another is to publicize British nuclear achievements and developments and to co-ordinate British participation in international conferences and exhibitions. No actual publications can be quoted at this stage but if the Forum develops along the lines of its American counterpart it could become one of the most useful sources of information on nuclear developments in the United Kingdom.

Other Organizations

There are a number of other organizations such as the Medical Research Council and the Agricultural Research Council which are concerned with certain specific aspects of nuclear energy and which publish information from time to time on developments within their field of study. Since, in most cases, the primary interests of these bodies are outside the nuclear field, their publications are described in the appropriate sections of Part 4.

CANADA

Atomic Energy Organization

In 1943, as a result of the war-time collaboration in atomic energy, a Canadian—British project was established at Chalk River, Ontario, with the object of designing and building a heavy-water-moderated reactor, the first reactor ever to be built outside the U.S.A. This partnership lasted until 1946 when the United Kingdom set up its own project. In the same year the Canadian Government passed the Atomic Energy Control Act by which a body known as the Atomic Energy Control Board was established to supervise the development and use of atomic energy in Canada.

As a result of further legislation, the Board's responsibility for nuclear research was delegated, in 1952, to a Crown Company, Atomic Energy of Canada Ltd., which assumed control of the research establishment at Chalk River, and another Crown Company, Eldorado Mining and Refining Ltd., was set up to act as a purchasing and marketing agency for uranium and uranium ores, raw materials in which Canada is particularly rich, reserves being estimated at nearly 400,000 tons of uranium oxide.

Whereas Britain has devoted most attention to the design of gas-cooled, graphite moderated reactors for power production, Canada has concentrated on reactors moderated and, in most cases, cooled by heavy water, using natural uranium as a fuel, and has developed both research and power reactors of this type. Four of these research reactors are located at the main research centre at Chalk River, about 130 miles from Ottawa, where the facilities include two particle accelerators and laboratories for physics, chemistry, metallurgy, electronics and engineering. Development studies are now being undertaken on the use of organic liquids as coolants in heavy-water-moderated reactors and an experimental reactor of this type is being built at the Whiteshell Nuclear Research Establishment near Winnipeg.

The research reactors at Chalk River produce considerable quantities of radioisotopes which are marketed by the Commercial Products Division in Ottawa, more than 90 per cent of the output being exported. The Division have specialized in the development of cobalt-60 beam therapy units for the treatment of cancer and have exported these units to nearly forty countries.

The Nuclear Power Plant Division of AECL, located at Toronto, was responsible for the design and construction of the prototype power reactor NPD-2 at Rolphton, Ontario, completed in 1962, and is now engaged with a full-scale version, known as CANDU, under construction at Douglas Point, Ontario. In both these projects AECL is co-operating with the Hydroelectric Power Commission of Ontario.

Canadian industry has taken an active part in atomic energy development not only in Canada but overseas where it played a major part in the design and construction of the Canada-India Reactor, a heavy water research reactor built at Trombay, India, as a joint project under the Colombo Plan. The main contractor for the two Canadian power stations is the Canadian General Electric Co., while fuel elements for the reactors are fabricated by AMF Atomics (Canada) Ltd. of Port Hope, Ontario. Another firm, Canadian Westinghouse Co., Hamilton, Ontario, is carrying out design studies of small enriched-uranium reactors for use in remote locations.

Information Sources

The information services of AECL are based on the Technical Information Office at Chalk River which is also responsible for the sale and distribution of

Canadian reports. AECL publish over 150 unclassified reports each year and these are deposited in the libraries of the National Research Council, Ottawa; McMaster University, Hamilton; and the University of British Colombia, Vancouver as well as being distributed to official projects and depository libraries in over thirty countries. In addition many papers are published in Canadian, British and American scientific journals and in the proceedings of national and international conferences. A list of reports and papers published up to 1959 has been issued [26] together with a number of supplements. Canadian reports are, of course, abstracted in *N.S.A.* and those for sale may be purchased from the Scientific Document Distribution Office, Atomic Energy of Canada Ltd., Chalk River, Ontario.

A good deal of information on nuclear research in Canada can be obtained from the annual reports [27] of Atomic Energy of Canada and from the press releases [28] which are put out from time to time. Outside the research field such matters as the control of nuclear materials and radioactive isotopes, reactor licensing, health and safety, technical training and international relations are the concern of the Atomic Energy Control Board and information on these aspects can be found in the Board's annual reports [29] to the Committee of the Privy Council on Scientific and Industrial Research.

The principal collections of foreign atomic energy reports in Canada are at Chalk River and in the National Research Council library at Ottawa. The National Research Council has long been associated with atomic energy; it was responsible for the Chalk River project from 1947 to 1952 and is still actively engaged in this field. It sponsors a number of scientific journals, including the *Canadian Journal of Physics* (see Chapter 8), which report some of the work being done in Canadian universities and other organizations in Canada.

AUSTRALIA

Atomic Energy Organization

An Atomic Energy Commission was established in Australia in 1953 under the general direction of the Minister of Supply although responsibility was later transferred to the Minister of State for National Development. At first the Commission was mainly concerned with the location and exploitation of uranium ores, of which there are extensive deposits in Australia, and it was not until 1956 that a research programme was started. The first research reactor, built by a British contractor, came into operation in 1958 at the AAEC Research Establishment at Lucas Heights, near Sydney.

Australia has large reserves of coal and there is considerable undeveloped hydroelectric potential, so that there is no immediate interest in a nuclear power programme. However, looking ahead, the Government have accepted the need to develop reactor systems which can provide economical power in certain areas of Australia by the late nineteen seventies and the design of such a reactor system is the main aim of the Lucas Heights establishment. Partly to avoid duplication of the research going on in other countries and partly to meet the particular needs of Australia, studies are being made of a high temperatur gas-cooled reactor in which the fuel is dispersed in a beryllium oxide moderator, the coolant being carbon dioxide.

Although Australia imports considerable quantities of radioisotopes for medical and industrial uses, there are some, e.g. short-lived isotopes, which have to be produced on the spot and this is done in the high flux reactor at Lucas Heights. Isotope production has increased markedly in the last two or three years and radioisotopes are now being exported to the Far East.

In order to encourage university participation in the atomic energy programme an organization known as the Australian Institute of Nuclear Science and Engineering has been established jointly by the AAEC and the Australian universities. The Institute arranges for scientists from the universities to carry out experiments using the research facilities at Lucas Heights, gives grants for research projects in the universities and sponsors short courses at individual universities on the use and applications of radioisotopes.

INFORMATION SOURCES

The AAEC publish a number of unclassified reports which are deposited in the Australian National Library at Canberra and the public libraries of New South Wales, Victoria, South Australia, Western Australia and Tasmania, and distributed to a number of overseas organizations including the UKAEA and the USAEC. Australian reports are abstracted in *N.S.A.* and listed in *Atomic Energy in Australia,* [30] a quarterly journal published by the AAEC, which contains original articles by scientists and engineers from research establishments, universities and other organizations and also gives details of the research programmes being undertaken in universities under grants awarded by the Australian Institute of Nuclear Science and Engineering.

The scientific research of the AAEC is reported in some detail in the Commission's annual reports [31] which also cover the work of the Australian Institute and include, each year, a concise but useful review of nuclear power developments in the major nuclear countries of the world.

Foremost among the universities in the nuclear field is the Nuclear Research Foundation within the University of Sydney which operates the School of Physics in the University. The Foundation publishes an annual report under the title *Nucleus* [32] which reviews each year's activities in the School.

INDIA

ATOMIC ENERGY ORGANIZATION

An Atomic Energy Commission was set up in 1948 as an advisory body only and reconstituted in 1958 as a policy making body with financial powers. Executive responsibility for atomic energy research and development resides with the Department of Atomic Energy which is responsible to the Prime Minister. The Department operates and finances a number of laboratories and institutes in India of which the most important is the Atomic Energy Establishment at Trombay in the Bombay area.

The Trombay establishment has developed rapidly into a major nuclear research centre employing over 5000 persons and is still growing. The facilities include three research reactors, one of which has been used since 1962 for the production of radioisotopes for agricultural, medical and industrial purposes in India, although its main use is as a materials testing reactor. The Establishment also operates a uranium extraction plant, a fuel element fabrication plant and reprocessing plant for irradiated fuel elements.

There are considerable deposits of uranium in India and the location and exploitation of these deposits is the responsibility of the Atomic Minerals Division which has a number of laboratories at Delhi engaged in mineralogical assay work and the design of radiometric surveying instruments. Although India has large coal reserves, many of the deposits are remote from industrial centres and the Government of India has placed considerable emphasis on nuclear power development in connection with its industrialization programmes. In addition to uranium ores, India has very large deposits of thorium in the form of monazite and is therefore interested in developing reactors using uranium-233 as a fuel. This is a fissionable isotope of uranium which does not occur in nature but which can be made by irradiating thorium-232 in a reactor. India's nuclear power programme, therefore, would be based on natural uranium in the beginning and later converted to uranium-233 if such a fuel cycle proved feasible.

At present one nuclear power station employing two boiling water reactors is under construction at Tarapur, the contractor being the General Electric Company of America and another of the CANDU type is planned for the Kotah area in Rajasthan.

The Department of Atomic Energy has encouraged nuclear research in the Indian universities by providing grants for specific research projects. In addition the Department wholly finances the Tata Institute of Fundamental Research in Bombay and the Physical Research Laboratory at Ahmedabad. The former is the national centre for fundamental research and its programme includes studies in nuclear and theoretical physics, plasma physics, radio astronomy and cosmic rays. The Ahmedabad Laboratory also carries out cosmic ray studies and both Institutes use the facilities of the Cosmic Ray Research Station at Gulmarg, in Kashmir, which is owned by the Department.

Information Sources

The information services of the Department of Atomic Energy are based on the Information Division of the Atomic Energy Establishment at Trombay. This is the main depository collection, in India, of foreign reports and journals in the atomic energy fields and the information so acquired is disseminated by means of a number of bulletins and digests which are compiled mainly for the use of the Establishment's staff but are sent to other institutions in India.

Information obtained as a result of research and development within the Department is published in the major scientific journals, particularly those of British and American origin, although some papers appear in the *Journal of Scientific and Industrial Research*. [33] A number of reports are issued by the Trombay establishment and some of these are distributed to foreign atomic energy projects and abstracted in *N.S.A.* Since 1962 the Department has published a monthly journal entitled *Nuclear India* [34] which describes recent atomic energy developments in the country.

By far the most useful sources of information on nuclear affairs in India is the annual report [35] of the Department of Atomic Energy which covers not only the work of the laboratories and divisions of the Department, but also that of the associated institutes. The report also gives details of papers and reports written by the Department's staff during the year under review.

PAKISTAN

An Atomic Energy Commission was appointed in 1956, responsible to an Atomic Energy Council presided over by the Minister for Fuel, Power and National Resources. Progress at first was disappointing, largely because of a lack of trained personnel, and steps were taken to select suitable people to be sent to universities and research establishments in Britain, the United States and other countries where they could receive the necessary training in nuclear science.

Pakistan is not plentifully supplied with conventional fuel resources and in some areas nuclear power could be produced more economically now than power from other sources. Pakistan is also interested in the application of radioisotopes and radiation sources for medical and agricultural purposes and takes part in the work of the CENTO Institute of Nuclear Science at Tehran, a research centre operated by the Central Treaty Organization of which Pakistan is a member.

Two training centres have been set up, one at Lahore and one at Dacca, where courses in the use of isotopes in medicine and agriculture are provided, and a new research centre, the Pakistan Institute of Nuclear Science and Technology, is being built at Islamabad. This will be equipped with a 5 MW research reactor, supplied by AMF Atomics with financial assistance from the United States, and the usual laboratory facilities.

Apart from a few reports little has yet been published about development in Pakistan but this situation will no doubt change when the Islamabad Centre becomes fully operational.

REFERENCES

1. CENTRAL OFFICE OF INFORMATION. *Nuclear energy in Britain*, 3rd ed. H.M.S.O., 1962.
2. UNITED KINGDOM ATOMIC ENERGY AUTHORITY. *The United Kingdom Atomic Energy Authority: its history and organisation*. UKAEA Public Relations Branch, 11 Charles II Street, London, S.W. 1. September, 1962 (DPR/INF/270). This publication also contains a list of the Acts of Parliament and other parliamentary documents relating to the organization of the Authority.
3. ATOMIC ENERGY RESEARCH ESTABLISHMENT. *Harwell handbook 1962: a brief guide to the Atomic Energy Research Establishment of the U.K.A.E.A.* Public Relations Section, AERE, Harwell, Berkshire, July 1962.
4. MINISTRY OF SUPPLY. *Harwell, the British Atomic Energy Research Establishment, 1946—1951*, by K. E. B. Jay. H.S.M.O., 1952.
    Includes a list of articles published by AERE staff during the period under review.
5. JAY, K. E. B. *Atomic energy research at Harwell*. Butterworth, 1955.
    Covers the period from 1952 to August 1954. Part 1 is a general survey of the main lines of research written primarily for the layman. Part 2 describes some of the more interesting fundamental studies and is written primarily for scientists.
6. JAY, K. E. B. *Britain's atomic factories*. H.M.S.O., 1954.
7. JAY, K. E. B. *Calder Hall: the story of Britain's first atomic power station*. Methuen, 1956.
8. UNITED KINGDOM ATOMIC ENERGY AUTHORITY. *Annual Report*. H.M.S.O. 1955 and onwards.,
    A popular version of the report is issued each year under the title "*Atom—*" (e.g. Atom, 1962) and sold by H.M.S.O.
9. *Atom* (Monthly). UKAEA, 11 Charles II Street, London S.W. 1. 1956 and onwards.
    A news journal describing new developments in the work of the Authority. It includes reprints of important lectures by senior staff of the Authority, press releases and extracts from parliamentary debates on atomic energy.
10. UNITED KINGDOM ATOMIC ENERGY AUTHORITY. *Guide to U.K.A.E.A. documents*. 3rd ed. Edited by J. Roland Smith. UKAEA, 11 Charles II Street, London, S.W.1. 1963.
11. INTERNATIONAL FEDERATION FOR DOCUMENTATION. *Universal decimal classification: special subject edition for nuclear science and technology; jointly sponsored by the International Atomic Energy Agency and the United Kingdom Atomic Energy Authority—(FID 351)*. International Federation for Documentation. 7 Hofweg, The Hague, Netherlands, 1964.
12. UNITED KINGDOM ATOMIC ENERGY AUTHORITY. *Film Catalogue 1962*. UKAEA, 11 Charles II Street, London, S.W.1.
13. *Scientific and technical news service (STN series)*. UKAEA, Public Relations Branch, 11 Charles II Street, London, S.W.1.
14. *Press Releases (AE series)*. UKAEA, Press Office, 11 Charles II Street, London, S.W.1.
15. UNITED KINGDOM ATOMIC ENERGY AUTHORITY. *List of publications available to the public*. (Monthly.) Library, Atomic Energy Research Establishment, Harwell, Berks. December 1955 and onwards.
    Covers (i)   Unclassified reports issued by Authority establishments.
    Covers (ii) Translations and bibliographies made and issued by Authority libraries.

(iii) Articles in periodicals written by Authority staff. The list also indicates those articles of which reprints may be obtained from Authority libraries.

(iv) Reports which have appeared in the published literature.

The annual cumulations are arranged under broad subject headings with author and report number indexes and are on sale through H.M.S.O.

16. ATOMIC ENERGY RESEARCH ESTABLISHMENT. *A list of reports and published papers by A.E.R.E. staff; papers published between 1952 and 1954*, by P. M. Harris and K. E. B. Jay. H.M.S.O., 1955 (AERE Inf/Bib 96).
The reports listed are those for sale only and cover the period up to 1954. For a list of articles by AERE staff written up to 1951 see reference 4.

17. ATOMIC ENERGY RESEARCH ESTABLISHMENT. *A list of reports for sale and published papers by A.E.R.E. staff, published mainly between August 1954 and December 1955*, by L. J. Anthony and P. M. Harris. H.M.S.O., 1957 (AERE Inf/Bib 96, supplement 1.)

18. ATOMIC ENERGY RESEARCH ESTABLISHMENT. *List of A.E.R.E. unclassified reports issued between 1952 and December 1955*, by L. J. Anthony and P. M. Harris. H.M.S.O., 1957 (AERE Inf/Bib 110).
These are reports which were not put on sale during the period.

19. *Atomic energy: Government publications sectional list no. 63.* Compiled by the Atomic Energy Research Establishment, Harwell. H.M.S.O., 1961.
Covers publications sold by H.M.S.O. up to 31 December 1960.

20. MICRO METHODS LTD. *Catalogue of published material available on micro-film, microfiche and microcard. Section E. Technical and science section.* Micro Methods Ltd., East Ardsley, Wakefield, Yorkshire. 1962.

21. NATIONAL INSTITUTE FOR RESEARCH IN NUCLEAR SCIENCE. *Annual report (1st) 1957—1958 and onwards.* Rutherford High Energy Laboratory, Harwell, Berkshire.

22. CENTRAL ELECTRICITY GENERATING BOARD. *Berkeley Nuclear Laboratories.* C.E.G.B., Public Relations Department, Friars House, Blackfriars Road, London, S.E.1. 1962.

23. CENTRAL ELECTRICITY GENERATING BOARD. *Annual report and accounts.* H.M.S.O. 1958 and onwards.

24. CENTRAL ELECTRICITY GENERATING BOARD. *Digest.* (Weekly.) CEGB, Information Services, Bankside House, Sumner Street, London, S.E.1.

25. UNITED KINGDOM ATOMIC ENERGY AUTHORITY. *The nuclear energy industry of the United Kingdom.* 2nd ed. rev. UKAEA, 11 Charles II Street, London, S.W.1. 1961.

26. ATOMIC ENERGY OF CANADA LTD. *List of publications. April 1952 to August 1959.* Atomic Energy of Canada Ltd., Scientific Documents Distribution Office, Chalk River, Ontario, 1959.
Supplements: 1. September 1959—May 1960. July 1960
2. June 1960—February 1961. March 1961
3. March 1961—November 1961. December 1961
4. December 1961—August 1962. September 1962

Reports and articles by AECL staff are listed under broad subject headings similar to those used in *N.S.A.*, and there are author and report number indexes.

27. ATOMIC ENERGY OF CANADA LTD. *Annual report*, 1952/53 and onwards. AECL, 150 Kent Street, Ottawa, Ontario.

28. ATOMIC ENERGY OF CANADA LTD. *Information for the press.* AECL, Public Relations Office, Chalk River, Ontario.

29. ATOMIC ENERGY CONTROL BOARD OF CANADA. *Annual report*, 1947/48 and onwards. Queens Printer, Ottawa, Ontario.

30. *Atomic Energy in Australian.* (Quarterly.) AAEC, 45 Beach Street, Coogee, N.S.W., Australia. December 1957 and onwards.
31. AUSTRALIAN ATOMIC ENERGY COMMISSION. *Annual report*, 1953 and onwards. AAEC, 45 Beach Street, Coogee, N.S.W., Australia.
32. *Nucleus.* (Annual.) Conpress Printing Ltd., 61–63 O'Riordan Street, Alexandria, Sydney, N.S.W., Australia. 1954 and onwards.
33. *Journal of Scientific and Industrial Research.* Parts A, B, C and D. (Monthly.) Council of Scientific and Industrial Research, Rafi Marg, New Delhi 1, India. 1942 and onwards.
    Section A is a review journal reporting on recent advances. Section B, C, and D contain original research papers.
34. *Nuclear India.* (Monthly.) Department of Atomic Energy, Apollo Pier Road, Bombay 1, India. September 1962 and onwards.
35. GOVERNMENT OF INDIA. DEPARTMENT OF ATOMIC ENERGY. *Annual report*, 1954 and onwards. Department of Atomic Energy, Apollo Pier Road, Bombay 1, India.

CHAPTER 3

INFORMATION SOURCES IN THE UNITED STATES

ATOMIC ENERGY RESEARCH IN THE U.S.A.

The history of atomic energy in the United States differs in at least one important respect from that in the United Kingdom, namely that industry and the universities have, from the beginning, played a very large part in research and development. Although the overall responsibility for the atomic energy programme is vested in the United States Atomic Energy Commission, research is carried out in about twenty major research centres which, although owned by the Commission, are operated by universities and private firms. In addition the Commission has placed research contracts with hundreds of smaller laboratories throughout the United States. In 1962 it was estimated [1] that altogether some 200,000 people were employed in jobs relating to the development and expansion of atomic energy in that country.

The most important of the major research centres are described below.

Ames Laboratory, Ames, Iowa

Operated by the Iowa State University of Science and Technology, it specializes in the metallurgy and chemistry of the lesser known elements and metals, and the development of ceramics and metal alloys for use in reactors.

Argonne National Laboratory, Lamont, Illinois

Established in 1946 and operated by the University of Chicago.

Research covers a wide field with emphasis on fundamental studies, high energy physics and reactor development. The Laboratory also houses the International Institute of Nuclear Science and Technology which is open to U.S. students and foreign nationals, and provides advanced training in nuclear science, particularly reactor design and engineering.

Bettis Atomic Power Laboratory, Pittsburgh, Pennsylvania

Established in 1948 and operated by Westinghouse Electric Corporation, its main job is the development and testing of reactors for marine propulsion.

Brookhaven National Laboratory, Upton, Long Island, N.Y.

Established in 1946 and operated by Associated Universities, an organization representing nine of the larger northern universities. Research covers most aspects of atomic energy particularly nuclear physics and technology, high energy physics, isotopes, effects of radiation, chemistry and metallurgy.

Hanford Laboratories, Richland, Washington

Operated by General Electric Co., this laboratory carries out research on reactor fuels, effects of irradiation on materials, reactor safety and waste disposal. Hanford is also the site of the Commission's main plutonium producing plant.

Knolls Atomic Power Laboratory, Schenectady, N.Y.

Established in 1946 and operated by the General Electric Co. Emphasis is on reactors for marine propulsion.

Lawrence Radiation Laboratory, Berkeley and Livermore, California

Both laboratories are operated by the University of California. Berkeley has developed from the original laboratory founded by Professor E. O. Lawrence in 1936. The laboratory specializes in high energy physics and it is here that all the transuranium elements, with the possible exception of nobelium, have been discovered. There are also extensive programmes of fusion research and isotope separation. Thermonuclear research is also being done at Livermore together with studies on propulsion reactors, nuclear weapons and the peaceful uses of nuclear explosions, a programme known somewhat euphemistically as "Plowshare".

Los Alamos Scientific Laboratory, Los Alamos, New Mexico

One of the oldest and largest of the Commission's laboratories, established in 1943 and operated by the University of California. In addition to weapons research the laboratory has extensive programmes in the fields of nuclear propulsion, fusion research and direct energy conversion, and is one of the two major centres of research into the peaceful uses of nuclear explosions.

National Reactor Testing Station, Idaho Falls, Idaho

Three contractors operate independently on this site, their programmes including reactor testing, development of breeder reactors, nuclear propulsion, materials testing and studies in reactor safety.

Oak Ridge National Laboratory, Oak Ridge, Tennessee

Operated by Union Carbide Nuclear Co. since 1948, it specializes in reactor research, isotope production (it is the largest supplier of radioactive and stable isotopes in the U.S.A.), chemical processing and plutonium separation, and thermonuclear research.

Sandia Laboratory, Sandia Base, Albuquerque, New Mexico

A weapons laboratory, operated by the Sandia Corporation, for developing the non-nuclear components of atomic weapons.

Savannah River Laboratory, Aiken, South Carolina

Operated by E. I. du Pont de Nemours and Co., research includes reactor physics and engineering and waste disposal. The laboratory is part of the Savannah River Plant the primary function of which is plutonium production.

An establishment of a different type is the Oak Ridge Institute of Nuclear Studies, a corporation of thirty-seven universities and colleges formed to enable the southern American universities to participate in research conducted at Oak Ridge National Laboratory. The Institute's own research activities are confined to the medical field, mainly in the application of radio-isotopes to the cure of cancer and other diseases. Training in the use of isotopes for medical and industrial purposes is given in the Special Training Division which has a function similar to that of the Isotope School operated by the U.K. Atomic Energy Authority at Wantage.

The Institute also houses the American Museum of Atomic Energy, opened in 1949, and has a large library in which the bulk of the material received by the AEC from all over the world as a result of exchange agreements is deposited.

A concise but comprehensive account of atomic energy research in the United States will be found in a booklet [2] entitled *Atomic Energy Facts* which describes the research programme of the AEC and its relations with industry. A particularly important current source of information is the annual report [3] to Congress of the AEC which summarizes the main developments in the research programme during the year under review.

USAEC INFORMATION SERVICES

In contrast to the research programmes, the technical information services of the AEC are largely centralized in the Division of Technical Information with its headquarters in Washington and its operations centre at Oak Ridge. One of the provisions of the Atomic Energy Act of 1954 was that information acquired as a result of research and development carried out by the AEC and its contractors should be made available to the scientific community and to others and the Division of Technical Information is charged with the task of carrying out this directive, not only by making information available but by doing everything possible to stimulate its use in industry and research.

Organization of the Division of Technical Information

The headquarters of the Division in Washington is responsible for the operation and planning of the technical information services and for developing standards, policies and procedures. To ensure that users are fully consulted the Commission has set up two advisory committees: the Technical Information Panel, established in 1948 and composed of representatives from the

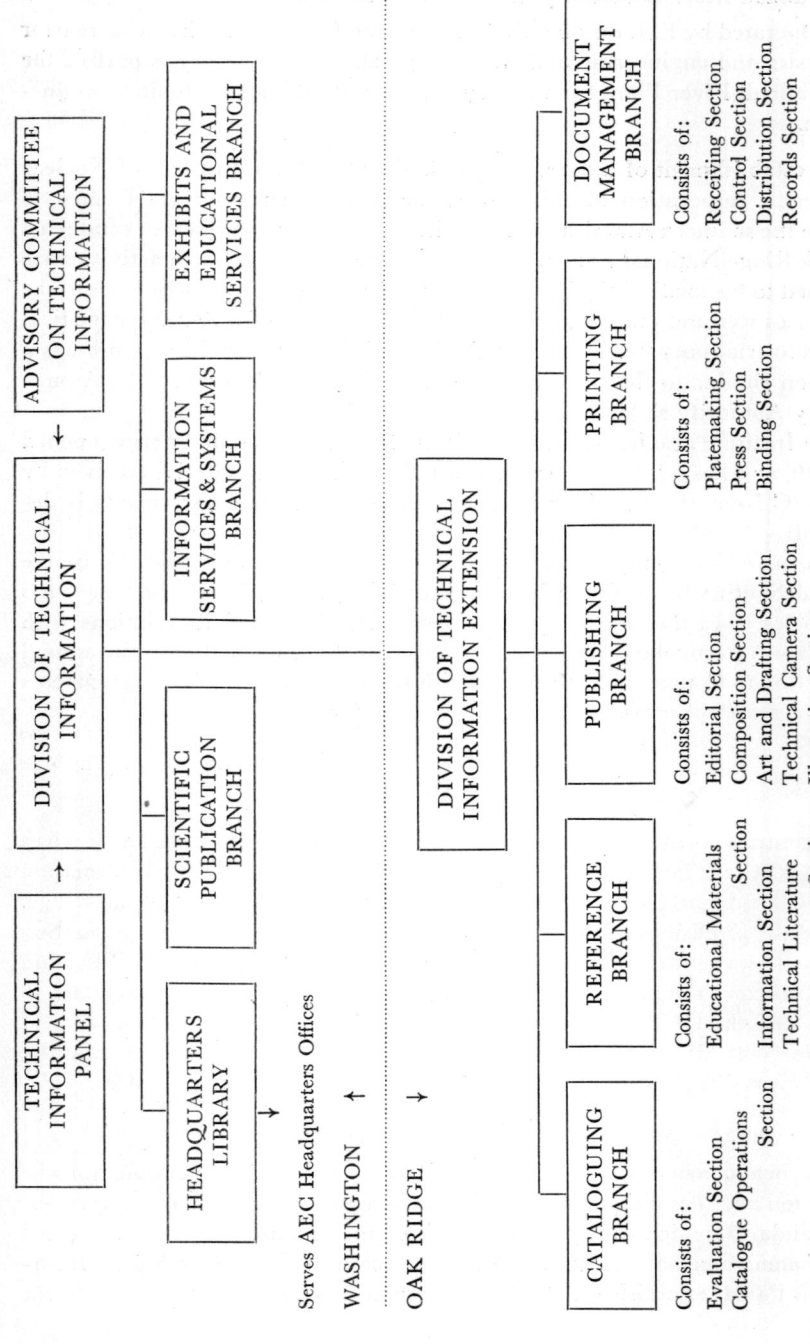

ORGANIZATION OF THE DIVISION OF TECHNICAL INFORMATION

AEC laboratories and major contractors, and the Advisory Committee on Technical Information, established in 1949 to assist in planning the Commission's industrial information programme. The members of the latter come mainly from organizations outside the AEC and include representatives from industry, the universities, commercial publishers and government departments.

Four main services are operated from Washington. These are:

The Headquarters Library which provides a service to the AEC headquarters staff and to those in the AEC Operations Offices in various parts of the country.

The Scientific Publications Branch which administers a programme for the writing and publishing of scientific textbooks, monographs, progress reviews and proceedings of conferences.

The Information Services and Systems Branch which collects and evaluates data on systems for acquiring, processing and disseminating technical information, advises the AEC and contractors on the installation of improved methods and serves as a focal point for liaison with other agencies on technical information matters.

The Exhibits and Educational Services Branch which plans and operates exhibitions and demonstrations aimed at students, teachers and the lay public. It also organizes technical information centres at exhibitions and conferences held outside the U.S.A.

The largest section of the Division and that most familiar to users of the Commission's information services is the Division of Technical Information Extension at Oak Ridge. DTIE is the major body for assembling and distributing information in support of the atomic energy programme in the United States and for organizing American participation in the field of international co-operation. It serves as a clearing house for AEC and non-AEC atomic energy literature and maintains the largest collection of atomic energy report literature in the country.

Services of the Division of Technical Information

Documentary resources

A large part of the information made available as the result of atomic energy research is issued in the form of technical reports. Those originating from the AEC and its contractors carry report numbers which have become standardized over a period of time and which should always be used in references and citations because each number constitutes a specific and unique identification of the report. It is true that in the early days of atomic energy, report numbers were being proliferated at a fantastic rate, and not only in America, and it was not unusual for distributing agencies to superimpose their own numbers on those of the originating organizations so that considerable confusion existed. However the position has improved greatly in recent years and serious attempts have been made to standardize procedures and to help users find their way through the maze of numbers which exist. A comprehensive list of report numbers used by the DTIE Cataloguing Branch will be found in the latest edition of TID–85 [4] which covers both AEC and non-AEC reports, including those issued by other countries.

A good deal of nuclear research is done by American agencies other than the AEC and the reports issued by these organizations are received at Oak Ridge together with those from foreign countries. British and Canadian reports have been sent to Oak Ridge from the early days of the projects and DTIE is now receiving regularly reports from more than twenty countries.

Since its inception the AEC has sponsored the publication of a number of textbooks, monographs and reviews. The earliest volumes, which were published by McGraw-Hill under the title *National Nuclear Energy Series*, summarized the information contained in early reports and papers, most of them produced during and just after the Second World War. Since then nearly one hundred technical books have been published in this way, generally through commercial publishers. All these volumes are listed in *Technical Books and Monographs* [5] which is obtainable from DTIE.

Information relating to scientific development in any particular field of atomic energy may be widely scattered in reports, journal articles, conference proceedings and other documents and it may be some time before this information is consolidated in monographs and textbooks. To help bridge this gap the AEC publish five quarterly progress reviews which digest and evaluate developments in selected fields. These are *Power Reactor Technology, Reactor Fuel Processing, Reactor Materials, Nuclear Safety* and *Isotopes and Radiation Technology*. The reviews are prepared independently by organizations under contract to the AEC and may be obtained on a subscription basis from the U.S. Government Printing Office.

Scientific journals are, of course, one of the most important means of communication between scientists and a good deal of information resulting from the Commission's research programme is made public in this form. DTIE receives copies of all journals which may contain articles of interest, including a large number received as a result of exchange agreements with foreign and domestic organizations. Most of these are deposited in the library of the Oak Ridge Institute of Nuclear Studies where they are available for loan to contractors and others in the United States. A list of organizations with which exchange agreements exist has been issued as TID–4554 [6] and gives titles of journals received together with their availability.

The AEC receives, every year, many hundreds of enquiries on atomic energy research from teachers and students at various levels. To deal with these enquiries the Commission have prepared a variety of educational materials in the form of booklets, charts, films and photographs from which an appropriate selection can be sent to each individual enquirer. Typical of these are the booklets in the *Understanding the Atom* series. [7] The information in these and similar booklets is supplemented by over one hundred reference lists, each dealing with a single topic, and referring in the main to publications which can be obtained fairly easily in public and other local libraries. In addition a large number of 16-mm films can be borrowed without charge by educational establishments in the United States and Canada and many of these are also obtainable outside the U.S.A. through the film libraries of the U.S. Information Service. These films are listed in two catalogues [8] which are obtainable from the AEC.

Dissemination of information

The most important guide to the vast body of literature which charts the progress of atomic energy research is *Nuclear Science Abstracts* which was originally issued by the AEC as a means of announcing the declassification and availability of its reports and is now the major abstracting journal in this field. Beginning in 1947 it was for a long time both a publicly available abstract journal and the AEC's accession list of unclassified material received. For this reason it frequently listed reports which, although unclassified, were not necessarily available to everyone who might ask for them. This was particularly true of non-AEC reports which were listed in *N.S.A.* under NP (i.e. non-project) numbers, often with no reference to the original report numbers. More recently the tendency has been to list only those items which can be readily obtained, either in printed form or as microcards or photocopies, and NP numbers are now used only for reports which do not bear originators' codes or which carry originators' numbers which are unacceptable for one reason or another.

Nuclear Science Abstracts (N.S.A.) is described more fully in Chapter 7. More detailed information about its scope and coverage will be found in TID–4552 [9] and those interested in the rules and techniques used by DTIE in compiling the journal should consult TID–4576. [10]

Up to July 1959, sets of catalogue cards for unclassified reports were distributed by the AEC to libraries inside and outside the United States and although these cards were of great value to other libraries in the atomic energy field, their sheer volume proved to be something of an embarrassment to both originators and recipients. In 1959 new techniques were introduced by which the indexes to *N.S.A.* were improved and by which cumulations could be promptly issued, so that the card service was no longer necessary and was therefore discontinued. The cards carried subject headings as well as the normal catalogue entry and abstract, and sufficient copies of each card were provided to enable users to maintain full subject catalogues if they so wished. These subject headings were the basis for those now used in *N.S.A.*; they are carefully controlled by DTIE and the full list is available to users in TID–5001 [11] which is revised periodically.

4 IAE

In addition to *N.S.A.* a large number of special bibliographies have been prepared and issued by the Division of Technical Information, both spontaneously and in response to specific requests. Most of these include abstracts or annotations of the publications referred to and cover not only report literature but journal articles, proceedings of conferences, books and other publications. They are usually issued in the TID report series and are indexed in *N.S.A.* and listed in TID–3043 and its supplements. [12] This list also gives details of bibliographies issued by non-AEC sources in America and other countries. The list is kept up to date by annual supplements and by another publication entitled *Bibliographies of Atomic Energy Literature* [13] which is issued at frequent intervals. A number of atomic energy projects co-operate with the AEC in providing materials for these lists which give details of newly available bibliographies completed and of bibliographies in preparation, thus helping to eliminate duplication of effort among the co-operating organizations.

Information about the current activities of the AEC is given in the press releases issued by the Commission and by the Joint Committee on Atomic Energy of the U.S. Congress. These are published in four series: (1) *Public Information Releases;* (2) *Speeches;* (3) *News Conferences;* and (4) *Industrial Information Releases;* and have been indexed back to 1947 in the *AEC News Release Index* [14, 15] issued by DTIE. Information as to the availability of individual press releases can be obtained from the Division of Public Information of the USAEC in Washington.

The AEC maintains a standard distribution system which provides for the direct distribution of reports by issuing laboratories to those AEC installations and other bodies which have a continuing need for this information. In addition it arranges that sufficient copies of unclassified reports shall be sent to Oak Ridge to distribute to depository libraries and foreign atomic energy projects with which the AEC has agreements. Reports are sent to nearly two hundred depository libraries in the U.S.A. and other countries and a full list of these will be found in any copy of *N.S.A.* In general the depository libraries will lend copies of the reports or provide photocopies at reasonable rates. Most of these reports are also put on sale through the Office of Technical Services of the U.S. Department of Commerce, Washington 25, D.C., from where they may be obtained either in printed form or as photocopies or microfilm. Reports sold by OTS, including those sold in facsimile or microfilm form, are listed in *Atomic Energy Commission Reports,* a semi-annual price list which can be obtained from OTS. Reports put on sale in the form of printed copies are also listed in the *Monthly Catalog of U.S. Government Publications* issued by the U.S. Government Printing Office. Microcopies of the reports can be purchased from Microcard Editions Inc., West Salem, Wisconsin, either by ordering individual copies or by placing a standing order for all reports as they are issued.

The U.S. Atomic Energy Commission is probably the largest single source of atomic energy information in the world and a full description of all the services it offers is outside the scope of this book. A comprehensive summary of the services provided to the public is contained in the latest edition of TID–4550 [16] which can be obtained from DTIE on request. A much fuller account, written primarily as a guide for AEC contractors, will be found in TID–485. [17]

OTHER SOURCES OF INFORMATION IN THE U.S.A.

A number of government agencies other than the AEC carry out or sponsor nuclear research in the United States. Chief among these are the Department of Defence, the National Aeronautics and Space Administration and the National Bureau of Standards.

The principal agency for collecting and disseminating defence reports is the Defence Documentation Center (DDC) which, up to July 1963, was known by the more familiar name of ASTIA (Armed Services Technical Information Agency). DDC published the *Technical Abstract Bulletin,* which is available to contractors and similar organizations, and the unclassified portion of this bulletin is reprinted in *U.S. Government Research Reports* which is described in Chapter 7.

Reports issued as a result of work sponsored by the National Aeronautics and Space Administration (NASA) are distributed by the Office of Scientific and Technical Information of NASA in Washington and are announced in *Scientific and Technical Aerospace Reports (STAR).* [18] Until late in 1963 these reports were also abstracted in *U.S. Government Research Reports* which was an excellent arrangement for those organizations who were interested only in the fringe areas of the aeronautical field. Unfortunately this arrangement no longer holds.

The National Bureau of Standards reports the results of its research in the *Journal of Research of the National Bureau of Standards* and in the *Technical News Bulletin.* A complete list of NBS publications, with abstracts, will be found in *NBS Circular 460* [19] and its supplements.

In general unclassified reports issued under federal auspices and made publicly available are sold through the Office of Technical Services. OTS has also established a number of regional technical report centres in the United States, each containing a collection of USAEC, DDC, and NASA unclassified reports as well as those of other government agencies. The centres, which provide reference and loan services and furnish photocopies, are located [20] in existing libraries strategically placed to give adequate coverage over the whole country.

Reports, which for one reason or another are not currently abstracted in *U.S. Government Research Reports,* can often be found in the *Monthly Catalog of United States Government Publications* which is particularly helpful in identifying non-technical documents such as those dealing with federal legislation in the atomic energy field. The Catalog is available by subscription from the U.S. Government Printing Office, Washington 25, D.C.

Acting as a link between the individual scientists and engineers working in the atomic energy field is the American Nuclear Society, a professional body founded in 1955 with headquarters in the John Crerar Library in Chicago. Physicists and engineers account for 80 per cent of the membership which is not confined to United States nationals. The Society holds national meetings twice a year, the proceedings of which are published in the *Transactions* (see Chapter 9), and issues a monthly news letter, *Nuclear News* (see Chapter 7).

One of the most useful sources of information on all aspects of nuclear energy, but particularly the commercial and industrial aspects, is Atomic Industrial Forum, a non-profit-making association formed in 1953 to encourage private participation in the industry. The association has both corporate and individual members, about a fifth of which are outside the United States. It holds an annual conference and a number of smaller meetings on specific topics and through a series of standing committees it studies and reports on special problems affecting the atomic energy industry in the United States. Typical of such reports are those on nuclear liability, insurance and indemnity, and the survey [21] on the future growth of the industry in America. It also publishes *Forum Memo*, [22] a monthly news summary which reports and interprets trends in the nuclear field. The Forum maintains a library at its New York headquarters and offers an information service to its members. It has the most comprehensive collection of atomic energy reports in America outside the AEC as it is a depository libary not only for AEC reports but for those issued by the main international and national projects throughout the world.

Before ending this account of American sources some mention should be made of the Joint Committee on Atomic Energy of the U.S. Congress. The Committee was originally established under the Atomic Energy Act of 1946, its terms of reference being "to make continuing studies of the activities of the U.S. Atomic Energy Commission and of problems relating to the development, use and control of atomic energy". It considers all legislation relating to atomic energy, monitors the working of the AEC and may propose policy changes in the national programme, but its most useful function is an informational one. Under Section 202 of the Atomic Energy Act of 1954 it is required to hold annual hearings to receive information concerning the devolpment and growth of the atomic energy industry and the transcripts of these "202" hearings, as they are called, are published and sold by the U.S. Government Printing Office. In addition the Committee may, from time to time, hold special hearings on particular topics, examples of which are *Indemnity and reactor safety*, *Radiation protection criteria and standards*, and *Technical aspects of detection and inspection controls of a nuclear weapons test ban*. In 1960 the

Committee sponsored a study of the *International atomic policies and programs of the United States* [23] which includes a good deal of information about developments in other countries. The transcripts of all these hearings are sold by the U.S.G.P.O. and constitute a most valuable source of detailed information on atomic energy matters in the United States.

REFERENCES

1. U.S. DEPARTMENT OF LABOR. BUREAU OF LABOR STATISTICS. *Employment opportunities in the atomic energy field.* Office of Technical Services, U.S. Department of Commerce, Washington 25, D.C., April 1962. (TID—14314.)
2. U.S. ATOMIC ENERGY COMMISSION. TECHNICAL INFORMATION SERVICE. *Atomic energy facts: a summary of atomic activities of interest to industry.* U.S. Government Printing Office, Washington 25, D.C., 1959.
3. (a) U.S. ATOMIC ENERGY COMMISSION. *Major activities in the atomic energy program.* (Annual.) U.S. Government Printing Office, Washington 25, D.C.
   (b) U.S. ATOMIC ENERGY COMMISSION. *Fundamental nuclear energy research.* (Annual.) U.S. Government Printing Office, Washington 25, D.C., 1962 and onwards.
4. U.S. ATOMIC ENERGY COMMISSION. DIVISION OF TECHNICAL INFORMATION EXTENSION. *Report number series used by the Division of Technical Information Extension.* USAEC, Division of Technical Information Extension, P.O. Box 62, Oak Ridge, Tennessee, 1964. (TID—85, 3rd rev.)
5. U.S. ATOMIC ENERGY COMMISSION. DIVISION OF TECHNICAL INFORMATION EXTENSION. *Technical books and monographs.* 4th ed. USAEC, Division of Technical Information Extension, P.O. Box 62, Oak Ridge, Tennessee, July 1963.
6. U.S. ATOMIC ENERGY COMMISSION. DIVISION OF TECHNICAL INFORMATION EXTENSION. *Organisations having publication exchange agreements with the AEC.* USAEC, Division of Technical Information Extension, P.O. Box 62, Oak Ridge, Tennessee, October 1961. (TID—4554, 1st rev.)
7. U.S. ATOMIC ENERGY COMMISSION. DIVISION OF TECHNICAL INFORMATION EXTENSION. *Understanding the atom: a series of booklets for students and laymen on various aspects of atomic energy and related fields.* USAEC, Division of Technical Information Extension, P.O. Box 62, Oak Ridge, Tennessee, 1963—.
   Booklets published to date are:

Our atomic world	Atoms in agriculture
Neutron activation analysis	Careers in atomic energy
Accelerators	Popular books on nuclear science
Nuclear reactors	Atomic fuel
Fall-out from nuclear tests	Direct conversion of energy
Power from isotopes	Controlled nuclear fusion

8. U.S. ATOMIC ENERGY COMMISSION. DIVISION OF PUBLIC INFORMATION, *Motion picture film library—popular level* and *Motion picture film library— professional level.* USAEC, Division of Public Information, Audio-visual Branch, Washington 25, D.C.
9. U.S. ATOMIC ENERGY COMMISSION. DIVISION OF TECHNICAL INFORMATION EXTENSION. *Subject scope of Nuclear Science Abstracts.* USAEC, Division of Technical Information Extension, P.O. Box 62, Oak Ridge, Tennessee. (TID—4552, 3rd rev. in preparation.)

Defines the subject fields of interest and the categories used in *N.S.A.* A statement on the subject scope of *N.S.A.* is also contained in *N.S.A.*, vol. 18, no. 1, January 15, 1964, Appendices 1 and 2.

10. U.S. ATOMIC ENERGY COMMISSION. DIVISION OF TECHNICAL INFORMATION EXTENSION. *Guide to abstracting and indexing for Nuclear Science Abstracts.* USAEC, Division of Technical Information Extension, P.O. Box 62, Oak Ridge, Tennessee. (TID—4576.)

11. U.S. ATOMIC ENERGY COMMISSION. DIVISION OF TECHNICAL INFORMATION EXTENSION. *Subject headings used by the USAEC Division of Technical Information.* Office of Technical Services, Washington 25, D.C., August 1962. (TID—5001, 4th rev.)

12. U.S. ATOMIC ENERGY COMMISSION. DIVISION OF TECHNICAL INFORMATION. *Bibliographies of interest to the atomic energy program, covering the period up to November, 1961.* Office of Technical Services, Washington 25, D.C., March 1962. (TID—3043, 2nd rev.)
First supplement, covering December 1961 to November 1962. March 1963. (TID—3043, 2nd rev. suppl. 1.)

13. U.S. ATOMIC ENERGY COMMISSION. DIVISION OF TECHNICAL INFORMATION. *Bibliographies of atomic energy literature issued or in progress.* USAEC, Division of Technical Information Extension, P.O. Box 62, Oak Ridge, Tennessee, 1963—.
Issued at two-monthly intervals as follows: December 1962 to January 1963 (TID—3715); February to March 1963 (TID—3716); April to May 1963 (TID—3717); June to July 1963 (TID—3718); August to September 1963 (TID—3719); October to November 1963 (TID—3720).

14. U.S. ATOMIC ENERGY COMMISSION. *Index to press releases of the Atomic Energy Commission. 1947 to 1951* (1953); *1952 to 1953* (1954); *1954 to 1955* (1956); U.S. Government Printing Office, Washington 25, D.C.

15. U.S. ATOMIC ENERGY COMMISSION. DIVISION OF TECHNICAL INFORMATION EXTENSION. *AEC News release index. January 1955 to December 1957* (1961); *January 1958 to December 1960* (1961); *January to December 1961* (1962); *January to June 1962* (1963); *July to December 1962* (1963). USAEC, Division of Public Information, Washington 25, D.C.

16. U.S. ATOMIC ENERGY COMMISSION. DIVISION OF TECHNICAL INFORMATION. *What's available in the atomic energy literature.* USAEC, Division of Technical Information Extension, P.O. Box 62, Oak Ridge, Tennessee, March 1963. (TID—4550, 9th rev.)

17. U.S. ATOMIC ENERGY COMMISSION. DIVISION OF TECHNICAL INFORMATION. *Technical information services of the USAEC.* USAEC, Division of Technical Information Extension, P.O. Box 62, Oak Ridge, Tennessee, November 1962. (TID—485, 4th rev.)

18. NATIONAL AERONAUTICS AND SPACE ADMINISTRATION. OFFICE OF SCIENTIFIC AND TECHNICAL INFORMATION. *Scientific and technical aerospace reports.* vol. 1, January 1963—. U.S. Government Printing Office, Washington 25, D.C. (Semi-monthly).
Covers reports issued by NASA and its contractors and by other agencies in the aeronautical and astronautical field and journal articles by NASA staff. Abstracts are arranged under thirty-four subject headings in order of NASA accession number and each issue has subject, corporate author, personal author, report number and accession number indexes. Cumulative indexes are published quarterly, half-yearly and annually. *STAR* is available on an exchange basis to universities, professional societies and independent research organizations.

19. NATIONAL BUREAU OF STANDARDS. *Publications of the National Bureau of Standards, 1901 to June 30, 1947*. U.S. Government Printing Office, Washington 25, D.C. 1948 (Circular 460).
   The following supplements have been issued:
   1. Supplementary list of publications of the National Bureau of Standards, July 1, 1947 to June 30, 1957. U.S. Government Printing Office, Washington 25, D.C. 1958. (Suppl. to Circular 460.)
   2. Publications of the National Bureau of Standards, July 1, 1957 to June 30, 1960: includes titles to papers published in outside journals, 1950 to 1959. U.S. Government Printing Office, Washington 25, D.C. 1961. (Miscellaneous Publication 240.)
   3. Supplementary list of publications of the National Bureau of Standards, July 1, 1960 to December 31, 1962. National Bureau of Standards, Washington, D.C. 1963.
20. OTS REGIONAL TECHNICAL REPORT CENTRES. University of California Library, Berkeley, California. University of California Library, Los Angeles, California. University of Colorado Library, Boulder, Colorado. Library of Congress, Washington, D.C. Georgia Institute of Technology, Atlanta, Georgia. John Crerar Library, Chicago, Illinois. Massachusetts Institute of Technology Library, Cambridge, Mass. Linda Hall Library, Kansas City, Missouri. Columbia University Engineering Library, New York. Carnegie Library of Pittsburgh, Pittsburgh, Pennsylvania. Southern Methodist University Science Library, Dallas, Texas. University of Washington Library, Seattle, Washington.
21. WARREN, F. H. AND OTHERS. *A growth survey of the atomic industry, 1958 to 1968*. Atomic Industrial Forum Inc., 860, Third Avenue, New York 22, N.Y. 1958.
22. *Forum Memo*. vol. 1, December 1953—. (Monthly.) Atomic Industrial Forum Inc., 850, Third Avenue, New York 22, N.Y.
   (Title changed to *Nuclear Industry* from vol. 11, no. 8, August 1964.)
23. JOINT COMMITTEE ON ATOMIC ENERGY. *Review of the international atomic policies and programs of the United States: report to the Joint Committee on Atomic Energy by Robert McKinney. Vol. 1, Report. Vols. 2—5, Background material*. U.S. Government Printing Office, Washington 25, D.C. October 1960.

CHAPTER 4

INFORMATION SOURCES IN THE SOVIET BLOC

NUCLEAR RESEARCH IN THE U.S.S.R.

It might be supposed that in a state in which administrative power is so firmly centralized, the organization of science and technology would take the form of a uniform structure in which each field of research was carefully demarcated and overlapping reduced to a minimum. Actually the organization of scientific research in Russia is, if anything, more complex than in any Western country, showing a diversity which gives ample opportunity to the individual scientists.

Nuclear research and development is undertaken by many different kinds of organization which can be broadly grouped under three headings:

(a) Institutes and laboratories of the U.S.S.R. Academy of Sciences and of the branch academies in the constituent republics of the Soviet Union.

(b) University institutes and technical colleges which are the responsibility of the Ministries of Technical Education in the nineteen republics of the U.S.S.R.

(c) Research centres run by the various specialized ministries which correspond in some respects to the private industrial research laboratories of the Western world.

On the civil side the work of these organizations is co-ordinated by the National Committee of the U.S.S.R. Council of Ministers for the Utilization of Nuclear Energy, while military applications are the responsibility of the Ministry of Medium Machine Building. The chain of command is somewhat complicated, however, because the scientific aspects of atomic energy research are co-ordinated by the U.S.S.R. Academy of Sciences in Moscow, which comprises a vast number of institutes, laboratories and departments with a membership, in 1959, of over 20,000 scientists. On the technological side the Ministry for the Construction of Electric Power Plants has responsibility for the nuclear power stations while other aspects, such as uranium production for example, are the concern of other Ministries.

It has been estimated that there are over 1500 establishments in the Soviet Union in which nuclear research or development of one kind or another is being undertaken and there is little reliable information on the activities of the great majority of these establishments. Probably the best known are those that specialized in nuclear physics before the Second World War. One of the most important is the Leningrad Physico-Technical Institute which has been doing research in nuclear physics since 1930 and is now engaged in fundamental research in solid state physics, nuclear physics, thermo-electricity, semiconductors and plasma physics. The Physico-Technical Institute of the Ukrainian Academy of Sciences at Kharkov has an equally long history. It was destroyed during the war but has been rebuilt with three laboratories, devoted to solid state physics, nuclear physics and theoretical physics respectively. The Institute also carries out research on high energy physics and controlled fusion. More recently established is the Kurchatov Institute for Atomic Energy in Moscow which has played a leading part in the development of atomic energy in Russia and is now a major centre for reactor development and plasma physics research. Other important laboratories include the Lebedev Institute in Moscow, the Institute for Electro-Physical Instruments in Leningrad, the Physical Institute at Kiev, and the Institute of Physics at Obninsk.

The most important laboratory devoted entirely to nuclear energy is the Joint Institute for Nuclear Research at Dubna, near Moscow, which was set up in 1956 as an international nuclear research organization within the Soviet Bloc. The states participating are Albania, Bulgaria, East Germany, China, North Korea, Mongolia, Poland, Czechoslovakia, Hungary, Roumania, North Vietnam and the Soviet Union, and the scientific programme of the Institute is controlled by a Scientific Council composed of three senior scientists from each member country. The Institute employs about 2500 persons, most of them Russian, and contains five major laboratories, two of them being concerned with high energy physics. A good deal of original research is carried on at the Institute, particularly by the Russian scientists, but from the point of view of the other participating countries its most important function is as a training ground, where their scientists can use equipment, such as the two large particle accelerators, which is not available in their own countries.

Information on atomic energy development in the Soviet Bloc is not easy to obtain and is often conflicting. This is hardly surprising since the release of information is carefully controlled and is designed to emphasize the successes and suppress the failures, so that a balanced picture is rarely obtainable. In spite of this two fairly comprehensive accounts [1, 2] of Soviet nuclear development have been published, based mainly on Russian sources and on the opinions of Western observers with first hand knowledge. Between them they give a fairly well balanced picture of Russian work in this field.

SOURCES OF RUSSIAN INFORMATION

The Academy of Sciences is the main scientific publishing body in Russia and much of the information on Soviet nuclear research which reaches the West does so through the medium of the Academy's journals. Apart from *Atomnaya*

Energiya none of these journals is devoted exclusively to nuclear energy; instead the subject is carried by articles in a large number of periodicals on physics, chemistry, geology, biology, medicine, technology and engineering. The most important of these are mentioned in Chapter 7.

A number of conferences, both national and international, have been held in the Soviet Union, each covering a particular aspect of nuclear science, and the proceedings of these are published by the Academy of Sciences. Russian scientists have also contributed extensively to international conferences held elsewhere, particularly the two United Nations conferences held in Geneva in 1955 and 1958.

In contrast to the situation in Britain and America there is no considerable programme of report publishing in Russia. Reports are issued by the various laboratories, in limited quantities and very few of these are sent outside the country. In many cases the reports are merely preprints of papers which will be issued in the scientific journals and the distribution of these is naturally restricted.

DISSEMINATION OF INFORMATION

In Russia there is no comprehensive abstract journal, such as *Nuclear Science Abstracts,* covering the whole field of nuclear literature. The All-Union Institute for Scientific and Technical Information (VINITI), which was set up in 1950 to centralize the processing of all technical literature in the U.S.S.R., issues *Referativnyi Zhurnal* [3] which is now published in 24 series representing the various individual branches of science and technology. This is the most comprehensive abstracting service in the world covering about 15,000 scientific and technical journals from over 100 countries in six languages. Over 700,000 abstracts are published each year of which over 100,000 are in the chemistry section and about 40,000 in the physics section, these being the most useful from the point of view of atomic energy. Abstracts normally appear about six months after publication of the original article, although in the case of Russian articles the delay is usually less than this.

Since it is Soviet policy to make foreign information available to Russian scientists as quickly as possible, the delay of six months in getting abstracts into *Referativnyi* is barely tolerable. To get over this difficulty the Institute publishes *Express Information* which contains summaries, in Russian, of the most important articles and patent specifications appearing in foreign journals. *Express Information* is published weekly in 61 series and the summaries appear one or two months after the original publication.

A new annual publication called *Achievements in Science* was started in 1964 with the object of recording progress in individual branches of science. This review journal contains references to sources and abstracts.

VINITI's prime responsibility is to provide information to Russian scientists and technologists as speedily and directly as possible. However it has been estimated that there are in the region of 25,000 technical libraries and information bureaux in the U.S.S.R. at the present time so that some duplication of effort is unavoidable. The possibility is being discussed of building up a unified state service of scientific and technical information on a subject specialization basis with VINITI playing a leading role as a co-ordinating centre for abstracting and for research in information handling techniques. With this in mind a decree has recently been issued which establishes the Universal Decimal Classification as the official subject classification for use throughout the Soviet Union. Such developments may well lead to some unification of atomic energy information which will make it easier for nuclear organizations in the West to keep track of Russian developments in this field, but for the present such organizations must make do with *Referativnyi* and with similar guides published outside the Soviet Union.

Of these *N.S.A.* devotes about 15 per cent of its contents to Russian literature, particularly the journals published by the Academy of Sciences. The titles of the Russian articles on nuclear energy which are abstracted in the various sections of *Referativnyi* have been collected since 1961 in *Ostliteratur*, compiled by the Atomic Energy Establishment at Julich in Western Germany, and appear from time to time in *Atomkernenergie Dokumentation Series C* published by the Gmelin Institute in Frankfurt. This journal is described in Part 4. From 1957 to 1959 selected abstracts from the Biology, Chemistry and Physics sections of *Referativnyi* were translated into English by the U.S. Joint Publications Research Service and issued as *Soviet Abstracts: Biology, Chemistry and Physics*. In 1961 these publications were superseded by *Abstracts of Selected Articles from Soviet Bloc and Mainland China Technical Journals* [4] issued by the office of Technical Services.

Availability of Soviet Literature

The volume of scientific publication in the Soviet Union has increased by over 500 per cent since 1950 and even with the aid of the abstracting and bibliographical services mentioned it is difficult to keep track of more than a small fraction of this output. The next step, i.e. that of locating a particular publication after a reference has been found, may, however, be even more difficult. One reason for this can be traced to the system of publication which operates in the U.S.S.R. In general books and periodicals are made available to the outside world through the agency of the International Book Centre which issues, to designated agents in various countries, lists of publications available. These official lists contain only about 50 per cent of the material actually published and their contents vary depending on the country to which they are sent, so that Russian publications obtainable in one country may be completely unavailable in another.

Usually the periodicals and books included in these lists are the more important ones such as the publications of the U.S.S.R. Academy of Sciences, and so a situation arises whereby those Western Libraries which make a serious effort to collect Russian literature all tend to acquire the same titles, and locations for journals not on the Book Centre's lists are exceedingly hard to find.

The largest collection of Russian literature in the West is that of the Library of Congress in Washington, which pursues an active exchange programme with Soviet establishments. Material thus acquired is listed each month in the *Monthly Index of Russian Accessions* [5] which is particularly good on technology but in which the basic scientific fields such as physics and chemistry are not so well represented. In Britain the most comprehensive collection is that of the National Lending Library for Science and Technology which now subscribes to about 1000 Russian journals, although this is only about 50 per cent of the number published. The NLL has issued two lists [6, 7] of the Russian journals which it holds; it also publishes an accessions list [8] of Russian books and monographs added to the collection.

One of the best collections of Soviet scientific and technical literature is that of the Technische Informationsbibliothek in Hanover, Western Germany, which subscribes to over 1400 Russian periodicals in all scientific fields except biology, medicine and agriculture.

In the nuclear field both the USAEC und UKAEA have extensive collections of those Russian journals which devote considerable space to the various aspects of atomic energy.

Books and journals which do not find their way into the lists issued by the International Book Centre are usually published only in sufficient quantities to satisfy the home market. They are listed in the *Soviet National Bibliography* but are usually out of print before their existence becomes known outside Russia. Some libraries, such as the NLL, have tried to get over this problem by placing "blind" orders with the official agents to supply everything published, but this policy can be no more than moderately successful because the Soviet publishing agencies are in many cases hard put to meet all the internal demands for scientific literature.

Report literature issued by the scientific institutes and laboratories is particularly difficult to obtain outside Russia and even libraries which have good exchange relations with Russian establishments find it impossible to obtain some report series even when their existence is known. In the nuclear field official exchange agreements exist between the National Committee for the Utilization of Nuclear Energy on the one hand and the USAEC, UKAEA and the French CEA on the other, as a result of which a number of Russian reports are received in these countries.

It is probable that the best way of obtaining many Russian publications is by direct exchange with the appropriate Russian institute or laboratory, particularly the All-Union Institute for Technical and Scientific Information. This body has recently been designated as the official depository for all scientific documents printed in limited numbers and has been made responsible for the offset duplication of this material.

Any exchange scheme must be based on adequate knowledge of the appropriate laboratories or organization in Russia and of the Soviet system for disseminating information internally. There is, in fact, an excellent description of this system in a translation, [9] issued by the Massachusetts Institute of Technology, of a Russian booklet originally entitled *Technological Information in Machine Building*. In spite of its title this is virtually the most comprehensive review of the Soviet technical information system to appear in Russia in recent years. A briefer, but more recent account, which describes the broad structure of the Soviet information system, is the report of the visit of a DSIR—Aslib team to Russia in June 1963. [10]

Awareness of what is being published followed by physical acquisition of publications is not the end of the story as far as Russian literature is concerned. Since only a negligible proportion of Western scientists can read Russian, most of this material has to be translated before it can be used. Many nuclear establishments in Western countries employ translators on their staffs, but the demand is such that only a small fraction of the Russian output can be dealt with in this way. In the last ten years considerable efforts have been made in Europe and America to make Russian literature widely available by intensive translation programmes, including cover-to-cover translations of the main Russian journals and conference proceedings, and the setting up of translation pools in various countries. Since translations are an essential part of the publicly available literature they are described more fully in Part 4.

OTHER SOVIET BLOC COUNTRIES

With the help of the Soviet Union a number of Eastern European countries are developing nuclear research and power programmes. Of these the most important are Czechoslovakia, Hungary, Poland and Yugoslavia, but nearly all have at least one research centre with a reactor purchased from the U.S.S.R.

Czechoslovakia has considerable reserves of uranium ore and is probably one of the most advanced of the satellite countries, with an ambitious nuclear power programme which includes the building of ten nuclear power stations of which the first is expected to come into operation in 1966. An Atomic Energy Commission was set up in 1955 with responsibility for nuclear power development and issues a monthly journal, *Jaderna Energie* [11] designed for scientific and technical staff in industry, research institutes and the universities. The main responsibility for nuclear research is carried by the Czechoslovak Academy of Sciences, which operates a national research centre, The Institute of Nuclear Physics, at Rez near Prague. Research is reported in the *Czechoslovak Journal of Physics*, [12] which is published in a domestic edition and in an international edition.

Although the birthplace of such well-known nuclear scientists as Szilard, Teller, von Neumann and Wigner, all of whom fled to America to escape Nazi oppression, Hungary had no atomic energy project until 1956 when an Atomic Energy Committee was established to plan and direct a research programme.

Some nuclear physics research had been carried out by the Central Institute of Physical Research of the Hungarian Academy of Sciences since 1950 and considerable deposits of uranium were discovered in 1954, but it was not until an agreement for co-operation had been signed with the Soviet Union in 1955 that atomic energy research got under way, based mainly on the Central Institute at Csilleberc, near Budapest, and the Institute of Nuclear Research at Debrecen. The Central Institute publishes a bimonthly journal, [13] available only by exchange, covering theoretical, experimental and nuclear physics and the Institute at Debrecen issues a quarterly journal [14] which, again, is devoted to theoretical and nuclear physics. The technological and engineering side of atomic energy in Hungary is mainly covered by *Energia es Atomtechnika*, a commercially published journal which is described in Chapter 7, but there is a considerable spread of information through a number of journals. Much of this can be located through the *Atomic Energy Bulletin* [15] issued by the Atomic Energy Committee, although its distribution is somewhat limited.

In Poland a Federal Commission for Nuclear Energy Affairs was established in 1956 to plan and administer the nuclear programme, with advice from the Committee on the Peaceful Uses of Atomic Energy of the Polish Academy of Sciences. Scientifically the research programme is directed by an Institute of Nuclear Research with three laboratories at Swierk and Zeran near Warsaw, and Bronowice near Cracow. The Institute issues a number of reports which are given wide distribution outside Poland, particularly to those countries with which there are exchange agreements. They are also sent to the USAEC in America and to the UKAEA and the National Lending Library in Britain. Much of Polish nuclear research is reported in the journals published by the Polish Academy of Sciences, particularly *Nukleonika* [16] which is published in English under a translation programme administered by the U.S. National Science Foundation, and contains translations of original Polish papers on nuclear technology, disposal of radioactive wastes, radiation chemistry and radiobiological protection.

In Yugoslavia an atomic energy project was established in 1955 under the Federal Commission for Nuclear Energy which has responsibility for nuclear raw materials, nuclear research, applications of atomic energy and radiological protection. There are three research centres, the Boris Kidrich Institute of Nuclear Research near Belgrade, the Rudjer Boskovic Institute, Zagreb and the Josef Stefan Institute, Ljubljana. The largest of these is the Boris Kidrich Institute which has three research reactors, one built by Yugoslavia, one supplied by the U.S.S.R. and one by the U.S.A. The Institute issues a quarterly *Bulletin* [17] with articles in English, and abstracts in English, Russian and French.

Some nuclear research is undertaken by Belgrade University and is reported in a technical journal [18] issued by the Faculty of Electrical Engineering of the University, and in other scientific publications. A useful guide to these is the *Bulletin of Documentation,* issued in five parts by the Yugoslav Centre for Technical and Scientific Documentation, of which the sections on electrical engineering and chemistry are the most useful. [19] A more detailed account of the development of atomic energy in Yugoslavia and the work of the major laboratories is contained in a book, [20] published in 1961, which includes a bibliography of articles, conference papers, books and pamphlets on atomic energy by Yugoslav authors.

In general the pattern of nuclear research in Eastern Europe is similar to that in the U.S.S.R., the most important bodies being the national academies of sciences and their associated institutes. Nuclear information can be found in a large number of journals, most of them published by the academies, and only exceptionally are reports issued or made publicly available. Most of the important journals are abstracted in *N.S.A.*, in the appropriate sections of *Referativnyi Zhurnal*, and in the domestic abstract journals where these exist. More detailed information can be obtained from the *Guide to the Scientific and Technical Literature of Eastern Europe* [21] a recent and very reliable guide published by the National Science Foundation, which lists all the major scientific periodical and bibliographical publications of these countries, with some useful notes on acquisition and availability.

REFERENCES

1. MODELSKI, G. A. *Atomic Energy in the Communist Bloc.* Melbourne University Press. 1959.
2. KRAMISH, A. *Atomic Energy in the Soviet Union.* Stanford University Press. 1959.
3. *Referativnyi Zhurnal.* All-Union Institute of Scientific and Technical Information (VINITI), Moscow, Lyubertsy, Oktyabr'skii Proezd, 403. 1953 and onwards. The main series of interest to nuclear scientists and engineers are as follows:
   Astronomy and Geodesy (Monthly); Biology (Fortnightly); Electrical and Power Engineering (Monthly); Automation and Electronics (Monthly); Physics (Monthly); Geophysics (Monthly); Chemistry (Fortnightly); Mechanical Engineering (Fortnightly).
4. *Abstracts of Selected Articles from Soviet Bloc and Mainland China Technical Journals.* (Monthly.) Office of Technical Services, Washington 25, D.C. 1961 and onwards.
   Series 1. Physics, geophysics, astrophysics, astronomy, astronautics and applied mathematics.
   Series 2. Chemistry, chemicals and chemical products.
   Series 3. Metallurgy, metal products and non-metallic minerals.
   Series 4. Engineering (aeronautical, civil, electrical, nuclear and structural) and machinery.
   Series 5. Communications, transportation, electrical and electronic equipment, systems and devices, aircraft and missiles.
   Series 6. General science, meteorology, biology, botany, zoology, medicine, fuel products and power.
5. LIBRARY OF CONGRESS. *Monthly index of Russian accessions.* U.S. Government Printing Office, Washington 25, D.C. 1948 and onwards.
6. DEPARTMENT OF SCIENTIFIC AND INDUSTRIAL RESEARCH. LENDING LIBRARY UNIT. *Titles of periodicals from the U.S.S.R. and "cover—to—cover" translations.* National Lending Library, Boston Spa, Yorkshire. March 1960.

7. DEPARTMENT OF SCIENTIFIC AND INDUSTRIAL RESEARCH. NATIONAL LENDING LIBRARY FOR SCIENCE AND TECHNOLOGY. *List of irregular serials received from the U.S.S.R. and Bulgaria.* National Lending Library, Boston Spa, Yorkshire. May 1961.
8. DEPARTMENT OF SCIENTIFIC AND INDUSTRIAL RESEARCH. NATIONAL LENDING LIBRARY FOR SCIENCE AND TECHNOLOGY. *List of books received from the U.S.S.R. and translated books.* (Monthly.) National Lending Library, Boston Spa, Yorkshire. 1959 and onwards.
9. MELIK-SHAKHNAZAROV, A. S. *Technical information in the U.S.S.R.*, translated from the Russian by B. I. Govokhoff. Massachusetts Institute of Technology Libraries, Cambridge, Massachusetts. 1961.
10. DEPARTMENT OF SCIENTIFIC AND INDUSTRIAL RESEARCH. *Scientific and technical information in the Soviet Union: report of the D.S.I.R.—Aslib delegation to Moscow and Leningrad, 7—24 June, 1963.* D.S.I.R. Information Division, State House, High Holborn, London, W.C.1. 1964.
11. *Jaderna Energie.* (Monthly.) Sponsored by Czechoslovak Commission for Atomic Energy, National Committee for Technical Development. SNTL—Publishers of Technical Literature, Spalena 51, Prague 1. 1955 and onwards.
    Covers construction of nuclear power stations, reactor technology, chemical processing and the application of radioisotopes. Articles in Czech and Slovak, headings and abstracts in English, Russian, French and German.
12. *Czechoslovak Journal of Physics.* (International edition.) (Monthly.) Postovní novinový úrad, Jindrisská 14, Prague 13. 1952 and onwards. This is section B of *Casopis pro Pestování Matematiky a Fysiky* (1872—1950). Covers nuclear research and instrumentation. Articles and abstracts in English, Russian and German.
13. *Magyar Tudományos Akadémia Központi Fizikai Kutató Intézetének Közleményei.* (Bimonthly.) Központi Fizikai Kutato Intézet, P.O.B. 49, Budapest 14. 1953 and onwards. Original articles in Hungarian with abstracts and headings in English and Russian.
14. *Atomki Közlemények.* (Quarterly.) M.T.A. Atommag Kutato Intézete, Debrecen, Hungary. 1959 and onwards. Articles and abstracts in Hungarian: table of contents in English. A supplement to vol. 5, no. 2, 1963 contains a complete list, in English, of publications by staff of the Institute up to December 1962.
15. *Atomtechnikai Tájékoztató.* (Monthly.) Kozponti Fiziki Kutató Intézet Dokumentációs Osztálya, P.O.B. 49, Budapest 114. 1958 and onwards. Translations and abstracts of review papers and special articles as well as bibliographies and lists of references in Hungarian.
16. *Nukleonika.* (Monthly.) State Scientific Publishing House, Miodowa 10, Warsaw. 1956 and onwards. (In English.)
17. *Bulletin of the Institute of Nuclear Sciences "Boris Kidrich".* (Quarterly.) Boris Kidrich Institute of Nuclear Sciences, P.O.B. 522, Belgrade. 1952 and onwards.
18. BELGRADE UNIVERSITY. FACULTY OF ELECTRICAL ENGINEERING. *Publications. Mathematical and Physical Series.* (Irregular.) 1956 and onwards. Covers nuclear reactors, radioactivity and electrical engineering.
19. *Bilten Dokumentacije, Elektrotehnika.* (Monthly.) Yugoslav Centre for Technical and Scientific Documentation, Belgrade. 1952 and onwards.
    Bilten Dokumentacije, Hemija i Hemiska Industrija. (Monthly.) Yugoslav Centre for Technical and Scientific Documentation, Belgrade. 1952 and onwards.
20. NAKIĆENOVIĆ, S. *Nuclear energy in Yugoslavia.* (In English.) Export Press, Dositejeva 21, Belgrade. 1961.
21. NATIONAL SCIENCE FOUNDATION. *A guide to the scientific and technical literature of Eastern Europe: Prepared for N.S.F. by Battelle Memorial Institute.* National Science Foundation, Washington 25, D.C. October 1962. (NSF—62—49.)

CHAPTER 5

INFORMATION SOURCES IN OTHER COUNTRIES

AT THE end of 1963, eighty-three countries were members of the International Atomic Energy Agency and therefore, it might be concluded, interested in some aspect of atomic energy. In many cases, however, the interest is confined to such aspects as the use of radioisotopes for medical and agricultural purposes, and only about half of these countries actually possess a nuclear reactor or have shown any indication of acquiring one.

The operation of even a moderate research and development programme in the atomic energy field requires considerable national resources and apart from those countries already mentioned, i.e. U.S.A., the British Commonwealth and the Soviet Bloc, not much more than a dozen countries have reached the stage where they can be judged to be contributing significantly to the common fund of knowledge of the subject.

In this chapter, therefore, only those countries are considered which have a comprehensive and well organized programme of research and development and which have taken steps to make the results of their work available in a readily accessible form.

FRANCE

The organization of atomic energy in France, a country which ranks fourth among the world's nuclear powers, is in many respects similar to that in Britain. An official body, the Commissariat à l'Énergie Atomique (CEA) is responsible for all aspects of nuclear research and development and co-operates with the state electricity generating organization, Électricité de France (EDF), in the design and construction of nuclear power stations, which are based on gas-cooled, graphite-moderated reactors as in Britain.

France is one of the five nations known to be engaged in the development of nuclear weapons, using plutonium derived from the three large production reactors at Marcoule, near Avignon. Small pilot plants for the production of uranium 235 by gaseous diffusion have been in operation since 1958 and a full scale diffusion plant is now under construction at Pierrelatte.

There are extensive deposits of uranium ore in metropolitan France and in the former French colonies in Africa, and France is one of the world's largest producers of uranium metal.

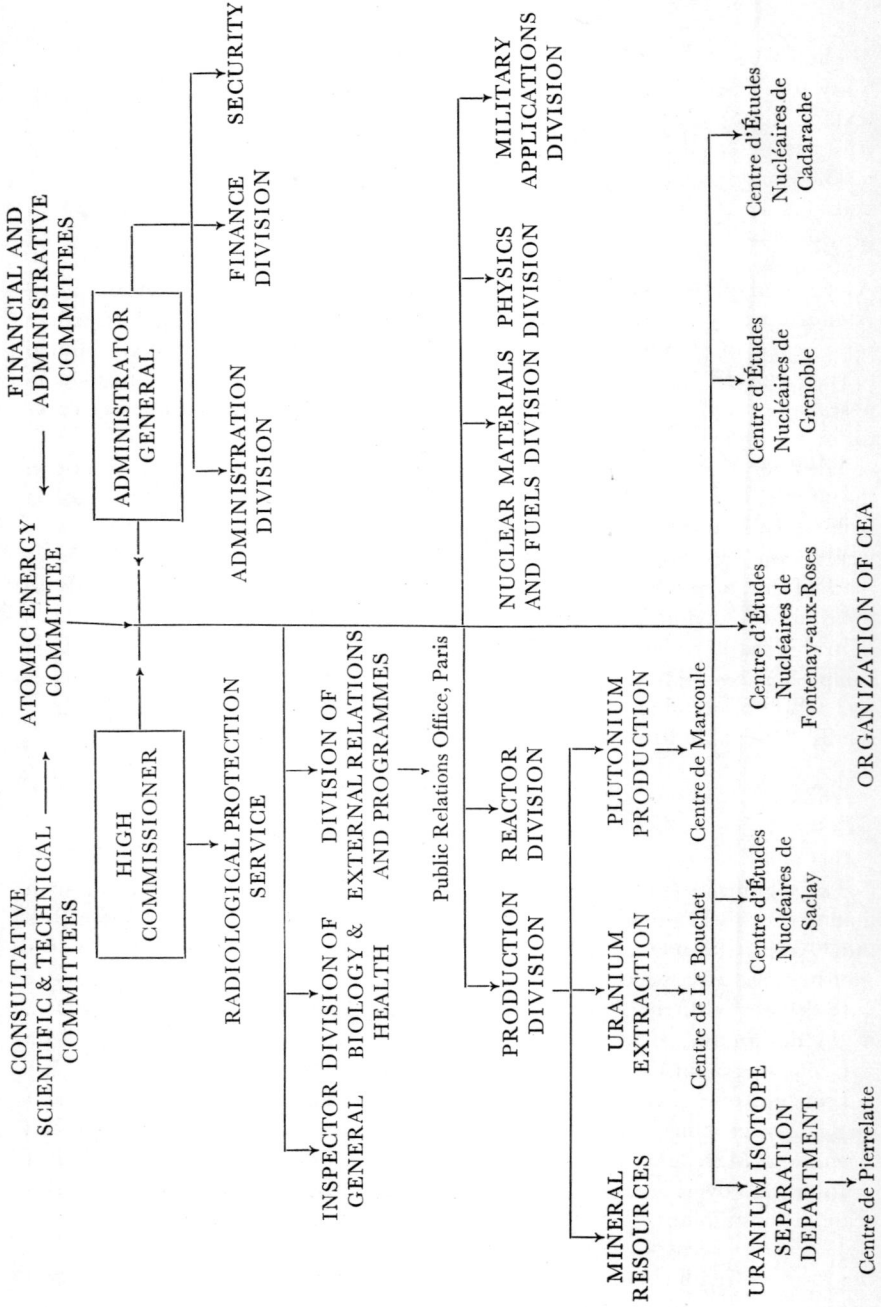

ORGANIZATION OF CEA

Commissariat a l'Énergie Atomique

The French Atomic Energy Commission was formed in 1945 with responsibility for developing the use of atomic energy in both civil and military contexts. Later, in 1947, it was given the additional task of developing the country's natural resources of uranium and other raw materials.

The Commission is responsible to the Prime Minister through an executive body known as the Comité de l'Énergie Atomique presided over by the Prime Minister. In practice the Administrator General, an official appointed by the Government, acts as chairman of the Committee and is also responsible for finance and administration. The scientific and technical programme is controlled by a High Commissioner, who is also a member of the Committee, and these two officials are responsible for the operation of the four main Technical Divisions. The research programme of the Divisions is spread over the four major research centres at Saclay, Fontenay, Grenoble and Cadarache.

Centre d'Études Nucléaires de Saclay. This laboratory, about twenty miles from Paris, is the largest of the research centres. Established in 1949 it now has a total population of about 8000 of which nearly 5000 are permanent staff, the rest being trainees and contract labour. The research programme covers all aspects of atomic energy including fundamental research, reactor physics and engineering, nuclear materials, fuel element fabrication, high energy physics, radiological protection, isotope separation and direct conversion. Saclay is also the home of the National Institute of Nuclear Science and Techniques, a training organization set up by CEA in conjunction with the universities to give advanced training in nuclear sciences. The Institute provides long and short courses on reactor engineering, radiobiology, radiochemistry, radiological protection and high energy physics.

Centre d'Études Nucléaires de Fontenay-aux-Roses. The construction of this laboratory began in 1955 on the site of France's earliest atomic laboratory at Fort de Chatillon, near Paris, where the first French reactor, Zoe, went critical in 1948. Research includes studies on radiological protection and the processing of nuclear fuels and raw materials. Fontenay is also the site of the controlled fusion project which is operated jointly by CEA and Euratom.

Centre d'Études Nucléaires de Grenoble. Opened in 1959, the laboratory has a broad programme of work covering reactor physics, nuclear physics, solid state physics, electronics and nuclear instrumentation, accelerator design, radiation chemistry and applications of radioisotopes. There is a close association with the University of Grenoble and, in fact, some of the laboratory's senior staff are also university professors.

Centre d'Études Nucléaires de Cadarache. This new laboratory, situated at Cadarache in the extreme south-east corner of France, is essentially a reactor research establishment with the emphasis on studies relating to power and propulsion reactors.

The work of these establishments is described in a series of booklets [1, 2, 3, 4] published by the Public Relations Service of CEA. A more general account of the research and development carried out by the CEA, in all fields, between 1945 and 1960 is described in a publication [5] issued in 1960, which also gives some information on the role played by French industry in the atomic energy field, and a list of the various decrees and ordinances governing nuclear developments in France. More recent information can be obtained from the annual reports [6] of the Commission and from the condensed versions [7] which are published in French and English.

CEA Information Service

The scientific and technical information service of CEA is based on the Central Documentation Service at Saclay, although there are, of course, information and library services at each of the major CEA establishments. The Documentation Service co-ordinates the work of the individual information services, edits and distributes scientific reports and papers, advises the individual libraries on techniques and methods, maintains a central collection of atomic energy literature and provides a library and information service for the scientific staff at Saclay.

The Commission issues reports and memoranda through the Central Documentation Service. Of these, reports are given a wide distribution to scientific and technical organizations in France and to foreign atomic energy projects. They are abstracted in *N.S.A.* and copies can be obtained either by exchange or, individually, by requests directed to the Service Central de Documentation, Centre d'Études Nucléaires de Saclay. Memoranda are produced in limited quantities only and are not given a wide distribution, but can be supplied in response to individual requests. Many CEA reports are published in scientific journals; these and reports that are not published in this way are all numbered in one numerical sequence, irrespective of the laboratory from which they originate. A list of reports [8] covering the period from 1948 to 1961 has been issued and this is kept up to date by a number of supplements.

Some of the research centres issue internal memoranda and reports which do not appear in the main CEA series. This is particularly true of Grenoble which has issued its own list of publications [9] covering reports, memoranda, patent applications and articles submitted to journals.

The *Bulletin d'Informations Scientifiques et Techniques* [10] which appears monthly is also edited by the Central Documentation Service and contains original articles on scientific research within the CEA together with annotated lists of recently published reports and translations. The Bulletin contains a section devoted to short notes, in both French and English, on recent events which have taken place in the atomic energy field in France. In general these are reprints of a separate publication, *Notes d'Information,* [11] which is issued more frequently, in French only.

Since September 1963, the CEA have also published a series of monthly *Rapports Economiques,* which are studies of the economic aspects of atomic energy, particularly the economics of reactor operation, fuel cycles, nuclear propulsion, etc.

Saclay is the main French depository library for world atomic literature and, in order to disseminate the information so received, the Documentation Service compiles and distributes *Bibliographie Scientifique Hebdomadaire*, a weekly list, of which each issue contains, on average, about 2000 entries. Each list entry is also prepared in card form for filing into the central information index which, by January 1962, had over two million entries with additions running at four hundred thousand per annum. The effort required to maintain this index by manual methods is so great that the CEA is now investigating the use of punched cards and computers and a special group has been set up for this purpose. One of the first results of this move is the publication of *Phys-index*, [12] a computer-compiled listing, in English, of periodical articles, with a keyword-in-context subject index, which is issued in four parts. About 150 scientific and technical periodicals are scanned for this index, very few of which are not already covered by N.S.A., and no abstracts or annotations are given.

A booklet [13] describing the work of the Central Documentation Service was issued in 1963. It lists the main series of publications of the Service and gives the addresses of the documentation units at the other CEA establishments.

The Public Relations Service of CEA, in Paris, arranges exhibitions and conferences and is responsible for a wide range of publications including the *Notes d'Information* and the booklets on the research centres already mentioned. The Service maintains an extensive photographic collection and has produced a number of films describing various aspects of the French programme. A catalogue [14] of these has been issued, to which supplements appear from time to time.

OTHER SOURCES OF INFORMATION IN FRANCE

Électricité de France (EDF)

The planning, construction and operation of nuclear power stations in France is the responsibility of Électricité de France, the public authority concerned with the generation and distribution of electrical energy. The first French nuclear power programme was adopted in 1955, and in 1957 work began on the power station at Chinon, on the Loire, where it was planned to construct three gas-cooled graphite-moderated reactors of a type similar to those of the British nuclear power stations. Of these the first two, EDF 1 and EDF 2, went critical in 1962 and 1963 respectively and the third, EDF 3, is still under construction. The building of the power station, and the design of the reactors, is described in a booklet issued by EDF in 1961. [15]

EDF have joined with the Belgian electrical utility company, Société Belge Centre et Sud, to form the Société d'Énergie Nucléaire Franco-Belge des

Ardennes, which is building another nuclear power station at Chooz near the Franco-Belgian frontier. The project is based on a pressurized water reactor expected to come into operation in 1965 and is being carried out under a U.S.A.–Euratom Joint Nuclear Power Agreement. More information about the work of Électricité de France in the nuclear field can be obtained from the annual *Rapport d'Activité*. [16]

Groupement Intersyndical de l'Industrie Nucléaire (GIIN)

GIIN is a grouping of trade associations charged with defining and defending the specific interests of nuclear industry in France. It operates through an Industrial Committee, which is the executive body, and six permanent commissions which have been set up to study and advise on questions of interest to the nuclear industry, particularly nuclear insurance and liability, marketing, industrial property, tariffs and external trade. Operating from its Paris headquarters at 64 avenue Marceau, Paris 8^e, its main function is to provide information on the nuclear services which the members of its component trade associations can provide and to represent, to public authorities and government departments, the interests of the nuclear industry in France.

*L'Association Technique pour la Production et l'Utilisation
de l'Énergie Nucléaire (ATEN)*

ATEN is the French equivalent of the British Nuclear Forum, although it was formed very much earlier (in 1955). Established with the support of CEA and EDF, its membership includes most of the French companies which have an interest in nuclear energy. It organizes conferences and exhibitions and provides a documentation service for its members, based on a comprehensive library of atomic energy literature located at its head office at 4 rue de Teheran, Paris 8^e. The library holds copies of all unclassified reports issued since 1946 by CEA, UKAEA and USAEC and maintains a subject index to these. ATEN publishes a bimonthly journal, [17] *Bulletin d'Information A.T.E.N.*, which contains general articles, summaries of nuclear developments in France and Western Europe, and announcements of events with which ATEN is connected.

Since April 1963, the Association has also published *Courrier de l'A.T.E.N.*, [18] a semimonthly journal designed to enable news of French nuclear developments to be disseminated more rapidly.

*Société Française pour la Gestion des Brevets d'Application Nucléaire
(BREVATOME)*

BREVATOME is a limited company of which 25 per cent of the shares are held by CEA, 5 per cent by EDF and the rest by industrial firms. Its purpose is to manage, on behalf of the owners, all French and foreign patents capable of nuclear application, belonging to the CEA, EDF and other shareholders. It grants licences, exchanges technical information and arranges the assignment of patents, and it participates in bids made by French industry for the supply of nuclear plant and equipment to foreign buyers. To carry out these functions BREVATOME has established, at its headquarters at 25 rue de Ponthieu, Paris 8e, a Patent Agency which prepares and files nuclear patents on behalf of its members, and a Documentation Office which collects patent specifications from all countries active in the nuclear field. To date about 20,000 specifications have been acquired, subject classified and indexed, and summaries of all patents received are published in its bimonthly journal, [19] *La Propriété Industrielle Nucléaire.*

Industrial Firms

French industry has taken an active part in nuclear development since the early days of atomic energy in France. There are a number of groupings, similar to the British Consortia, which are now in a position to design, construct and put into service complete nuclear plants, including power stations, and French industry is keenly conscious of the export value of such capability. The best guide to the services offered by these firms is the *Annuaire de l'Activité Nucléaire Française* [20] which is produced jointly by GIIN and ATEN.

ITALY

Italy is a country with negligible coal resources, a little oil, and considerable hydroelectric potential which is being brought into use as quickly as possible. It is estimated, however, that all economic hydroelectric sources will be harnessed by 1965, so that it is hardly surprising that a vigorous nuclear power programme is being pursued. Efforts in this direction are fairly evenly divided between state controlled organizations and private enterprise, although in Italy it is not always easy to distinguish one from the other.

Comitato Nazionale per l'Energia Nucleare

The first Italian official organization for atomic energy, Comitato Nazionale per le Richerche Nucleare (CNRN), was formed in 1952 to organize and encourage nuclear research and development in Italy. It operated under the general direction of the National Research Council and of the Minister of Industry and Commerce and in spite of inadequate funds it laid a firm groundwork for nuclear development in Italy. In 1960 this body was reconstituted as Comitato Nazionale per l'Energia Nucleare (CNEN), a public authority under the direct control of the Minister of Industry and Trade, with responsibility for nuclear research, the control of nuclear materials and the co-ordination of atomic energy development in Italy.

CNEN operates three major research centres at Casaccia, Frascati and Bologna respectively and has a considerable interest in the Euratom research centre at Ispra.

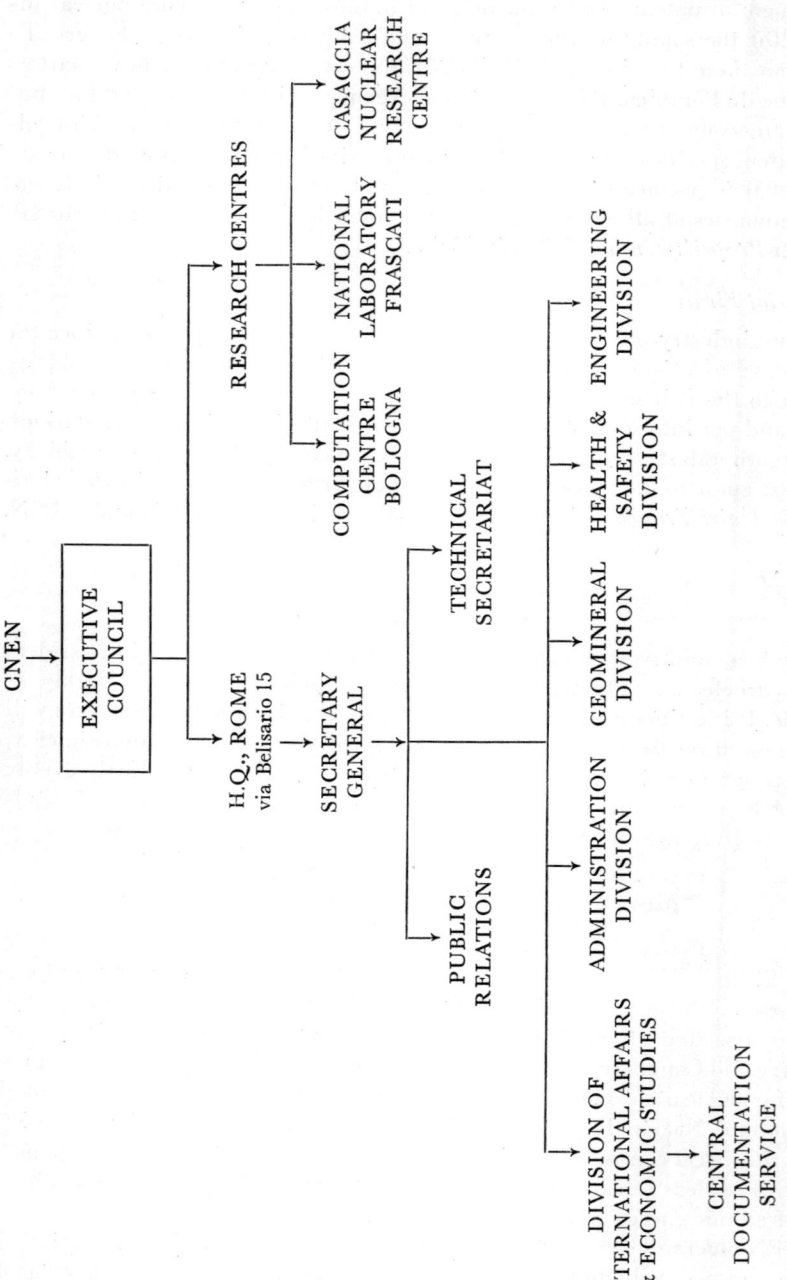

ORGANIZATION OF CNEN

Casaccia Centre for Nuclear Studies. This establishment, twenty-three miles north of Rome, is the principal reactor development laboratory and is operated directly by CNEN. The main objective of the Laboratory is the development of experimental reactors using organic moderators, which will provide information for the subsequent design of power reactors of this type. Other studies include research into methods of geochemical and mineralogical prospecting and the use of isotopes and radiations in agriculture.

Frascati National Laboratory. Frascati, twelve miles south of Rome, is a centre for fundamental research in nuclear physics and high energy physics, operated, on behalf of CNEN, by the National Institute of Nuclear Physics, a body formed originally by the fusion of the Physics Institutes of the universities of Rome, Padua and Turin. The Ionized Gas Laboratory, which is the Italian centre for plasma physics research, is also located at Frascati.

Computation Centre, Bologna. This centre has been established recently to provide computer and mathematical services for the nuclear laboratories in Italy. It is also carrying out theoretical studies on fast reactors.

Centre for Nuclear Studies, Ispra. The Ispra Centre was built in 1957–1958 by CNRN, as the main nuclear research laboratory in Italy. It was, however, taken over by Euratom in 1960, the agreement providing that CNEN could continue to carry out a research programme at the Laboratory. The work is conducted by Nucleare Italiana, a government-owned corporation set up by CNEN for this purpose.

The first report [21] issued by CNRN covered the years 1952–1957 and was used as the basis for its first five-year development programme. This was followed by an activities report [22] for 1958–1959 and then by the annual reports [23] of CNEN which give the most recent information on nuclear research in Italy. The work of CNRN between 1952 and 1959 is also described in a booklet [24] published just before the Committee was reconstituted and similar booklets have been issued for the main research centres. [25, 26, 27]

INFORMATION SERVICES OF CNEN

The main documentation centre of CNEN, which is the principal Italian depository for foreign reports, is housed in the headquarters in Rome and works in close co-operation with the documentation centres in the various laboratories. Reports are issued by all the establishments, usually in two series; technical reports, which are given a wide distribution and abstracted in *N.S.A.*, although the majority of them are reprints of articles submitted to scientific and technical journals, and internal reports which have a more limited circulation, although many of these are either preprints or reprints. CNEN has published catalogues [28, 29] of reports which include those issued by the Casaccia and Ispra centres but not those issued by Frascati. The latter have been listed separately [30] and supplements to this list appear in the half-yearly progress reports of the Frascati National Laboratory. [31] CNEN also publishes a monthly journal, *Notiziario*, [32] which surveys developments in Italy and abroad and which contains, from time to time, up-to-date lists of reports and journal articles published by CNEN.

OTHER SOURCES OF INFORMATION IN ITALY

Centro Informazioni Studi Esperienze (CISE)

Founded in 1946, this is the oldest Italian organization active in the atomic energy field. It is supported equally by public and private funds and owns extensive laboratories near Milan devoted mainly to reactor development studies, solid state physics and radiochemistry. Research results are published in scientific journals and in the form of reports which are widely distributed and abstracted in *N.S.A.* The CISE Documentation Service receives reports from all the major foreign projects and has an extensive library of nuclear literature.

The work of the CISE laboratories is described in a publication [33] issued in 1959, which contains a list of reports and articles issued up to that time. A more detailed list, covering the period 1946 to 1960, has also been issued [34] and this is kept up-to-date by annual supplements. CISE also publish *Energia Nucleare*, a monthly journal which is described in Chapter 7.

Forum Italiano dell'Energia Nucleare (FIEN)

The Italian equivalent of the Atomic Industrial Forum, it offers services similar to those of its American counterpart. It is concerned mainly with the economic and industrial aspects of nuclear energy and has arranged a number of conferences and meetings in this field. It publishes *Atomo e Industria*, [35] a semimonthly journal which carries articles and news items in English and Italian on international and national developments, particularly in the industrial field.

Associazione Nazionale di Ingegneria Nucleare (ANDIN)

This is an association, mainly of engineers, for the study of practical problems connected with the construction of nuclear installations. It publishes, with the support of CNEN, a bimonthly journal, *Rivista di Ingegneria Nucleare*, [36] which contains original articles on nuclear engineering and technology as well as a section devoted to abstracts of articles from nuclear engineering journals published throughout the world.

Industrial Sources

Industry's main contribution to nuclear development in Italy has been in the field of applied nuclear science and particularly in the construction and operation of the nuclear power stations. The first one at Latina was built by the British Nuclear Power Plant Company in co-operation with AGIP Nucleare which was a subsidiary of Ente Nazionale Idocarburi (ENI), a government-owned holding company responsible for the exploitation of natural resources in Italy, and is now a division of Societa per Azioni con Sede in Milano (SNAM). The work of AGIP Nucleare is described in the annual reports of ENI, which are published in English and Italian. AGIP Nucleare has research laboratories at San Donato, near Milan, and owns 75 per cent of the Societa Italiana Meridionale Energia Atomica (SIMEA), a company set up to operate the Latina power station. The other 25 per cent is owned by FINELLETTRICA, a public authority, with headquarters in Rome, which manages the public section of Italy's electrical utility.

The Societa Richerche Impianti Nucleari (SORIN), Milan, a joint subsidiary of the great industrial combines FIAT and Montecatini, has research laboratories at Saluggia near Turin and is engaged with AGIP Nucleare in building, under the direction of CNEN, a prototype power reactor, using an organic moderator, at Lake Brasimore near Bologna. SORIN issues technical reports which are sent to atomic energy projects in the major nuclear countries and are abstracted in *N.S.A.*

JAPAN

Electric power consumption in Japan has risen rapidly since the end of the war and most of the easily exploited sources of coal and water power have been brought into use. There is, therefore, a need for alternative economic energy sources and a full-scale nuclear power programme has been put into operation. The first nuclear power station is now under construction and a target of 1000 MW of nuclear capacity by 1970 has been set in the first instance.

Atomic Energy Organization

An Atomic Energy Commission was created in 1956 as an advisory body to the Prime Minister, with responsibility for developing a national policy for atomic energy and co-ordinating research and experiment in the nuclear field. An executive body, the Atomic Energy Bureau of the Science and Technics Agency, acts as secretariat to the Commission and is responsible for administering and financing nuclear research and development, controlling nuclear materials, training, health and safety, promoting the use of radioisotopes and, more specifically, for the administration of the three atomic energy development organizations, the Japan Atomic Energy Research Institute, the National Institute of Radiological Sciences and the Atomic Fuel Corporation.

Japan Atomic Energy Research Institute (JAERI)

The Institute was created in 1956 as a special corporation for which 90 per cent of the funds were provided by the Government. It has a headquarters in Tokyo and two laboratories at Tokai-mura and Takasaki respectively.

66　NATIONAL SOURCES OF INFORMATION

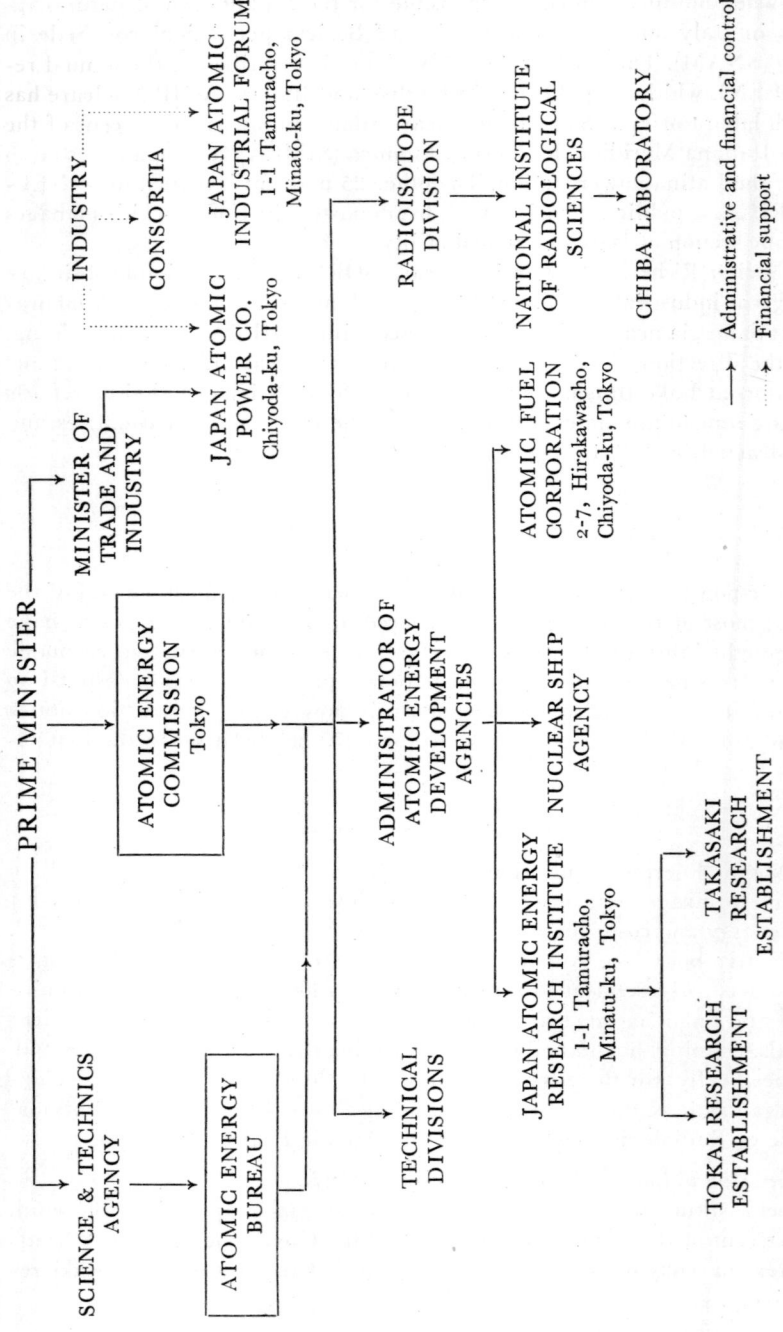

ATOMIC ENERGY ORGANIZATION IN JAPAN

Tokai Research Establishment. This is the oldest and largest of the laboratories, situated at Tokai-mura, about 75 miles north of Tokyo. The laboratory has four research reactors and a prototype power reactor in operation and maintains an Isotope School and a Reactor Training School.

Takasaki Research Establishment. This is a new laboratory, still under construction, which will be devoted to radiation chemistry and associated studies. The facilities will include a particle accelerator and a powerful cobalt-60 irradiation facility.

National Institute of Radiological Sciences (NIRS)

This Institute, established in 1957, is responsible for research into the medical uses of radiation and the effects of radiation on living organisms. Its main laboratory is at Chiba, a few miles from Tokyo.

Atomic Fuel Corporation

This is a public corporation, set up in 1956, to develop natural resources of uranium and other nuclear materials and to be responsible for the fabrication and reprocessing of nuclear fuels. It has a metal refining plant at Tokai-mura, adjacent to the Tokai Research Establishment.

The organization of the Atomic Energy Bureau and its associated Institutes is described briefly in a booklet [37] issued by the Commission in 1962. A more detailed account [38] of the first long-range programme for atomic energy development in Japan was issued in February 1961 and a more recent description [39] of the work of the Tokai Research Establishment has been made available.

Information Services

Atomic Energy Commission

Japan is the only country, outside the U.S.A., which publishes an abstract journal devoted exclusively to all aspects of nuclear science. This is *Nuclear Science Abstracts from Japan* [40] published in English by the Japan Atomic Energy Research Institute in order to bring the results of Japanese research to the attention of Western scientists. The journal has been published in a revised form since October 1963 and now consists of abstracts for reports and articles in scientific and technical journals, published in Japan, which contain research results or surveys of important developments in Japanese nuclear science and technology. Copies of these abstracts are sent to the USAEC and some of them appear in *N.S.A.* This journal, together with annual reports and technical reports, is edited and issued by the Division of Technical Information at the Tokai Research Establishment, where the main depository collection of foreign and domestic atomic energy publications is maintained. Research reports, generally in Japanese, with English summaries, are distributed to the larger foreign atomic energy projects and listed in *N.S.A.* The Institute also issues a monthly journal in Japanese [41] which contains titles and abstracts of reports by JAERI staff.

Japan Atomic Industrial Forum (JAIF)

This organization, which was established in 1956, is one of the most prolific sources of information on atomic energy affairs in Japan. Apart from sponsoring a number of conferences, exhibitions, symposia and special meetings, it has organized teams to visit and report on foreign nuclear installations and is actively co-operating with a number of foreign agencies. Most important are its information activities which include the publication of eight periodicals of which one is published in English [42] and given a wide distribution overseas.

The Forum issues an *Atomic Energy Yearbook* [43] in Japanese and has published an *Atomic Energy Directory of Japan* [44] in English, which covers every type of organization having an interest in atomic energy in Japan, although it is now getting a little out-of-date.

Japan Atomic Power Company and the Japanese Nuclear Industry

The Japan Atomic Power Company is a corporation set up to finance the building of nuclear power stations, starting with the one at Tokai-mura. The Government owns 18 per cent of the shares, the rest being spread among the public and private electrical utility companies and private industry. An account of the work of the Company, which includes a description of the Tokai power station, is contained in a booklet [45] issued in 1961. The power station is being built by the British General Electric Company in association with the First Atomic Power Industry Group, one of the five atomic energy consortia in Japan. All five of the consortia have research laboratories, most of them with research reactors, and are rapidly building up considerable experience in the manufacture and assembly of reactor components. A detailed description of their activities is given in a special number of *Atoms in Japan* [46] issued in 1963.

Societies and Universities

The Atomic Energy Society of Japan is an association of scientists and engineers working in the atomic energy field. It publishes a monthly journal [47] which contains original articles in English and Japanese as well as surveys of particular aspects of Japanese nuclear research. Many of its members are university scientists since most of the Japanese universities are engaged in research on some aspect of nuclear science.

The Institute for Nuclear Study of the University of Tokyo is the main centre for high energy physics research, with two accelerators which are available for use by nuclear physicists throughout the country. The Institute issues an annual report [48] which gives a detailed account of the research carried out and includes a list of reports and journals articles published during the period under review. Reports are usually in English and are widely distributed and abstracted in *N.S.A.*

Other important laboratories are at the Universities of Kyoto, Kyushu, Nagoya, Osaka and Tokai and there is a Nuclear Research Laboratory at the Tokyo Institute of Technology.

FEDERAL REPUBLIC OF GERMANY

The first observation of the fission of uranium was made by the German chemists Hahn and Strassman [49] in 1938 at the Kaiser Wilhelm Institut für Chemie in Berlin, and although they were unable to accept the full implications of their results, these were correctly interpreted by another German scientist Lise Meitner, who had fled from Nazi Germany to Stockholm, and her Austrian-born nephew Otto Frisch.

Unfortunately from Germany's point of view, further experimentation in that country was cut short by the war and the scientists who fled to Britain and America took their knowledge with them. It was not until 1955 that Germany once more entered the atomic energy field and the first steps were taken to plan a programme of nuclear research by setting up a Ministry for Atomic Energy and Water Resources. In the following year an Atomic Energy Commission was formed to advise the Minister on nuclear research and development.

In 1962 the Ministry for Atomic Energy was replaced by a Ministry for Scientific Research covering all aspects of scientific research and particularly atomic energy and space research, and in 1963 the nuclear research and power programmes were recast and intensified.

Western Germany now has eighteen nuclear reactors in operation or under construction and although many of these are the work of foreign contractors, a good deal of experience has been gained and the more recent reactors have been designed and built entirely by German industry, which has set up a number of consortia for this purpose.

Atomic Energy Organization

The Minister for Scientific Research is responsible for co-ordinating atomic energy research in Germany, and is chairman of the Deutsche Atomkommission which formulates policy. Although the Atomkommission has no executive power it guides the research effort through six standing committees and a large number of working parties which between them cover every conceivable aspect of the subject.

Reporting also to the Minister is the Reaktor-Sicherheitskommission (Reactor Safety Committee) which examines all proposals for the building of reac-

NATIONAL SOURCES OF INFORMATION

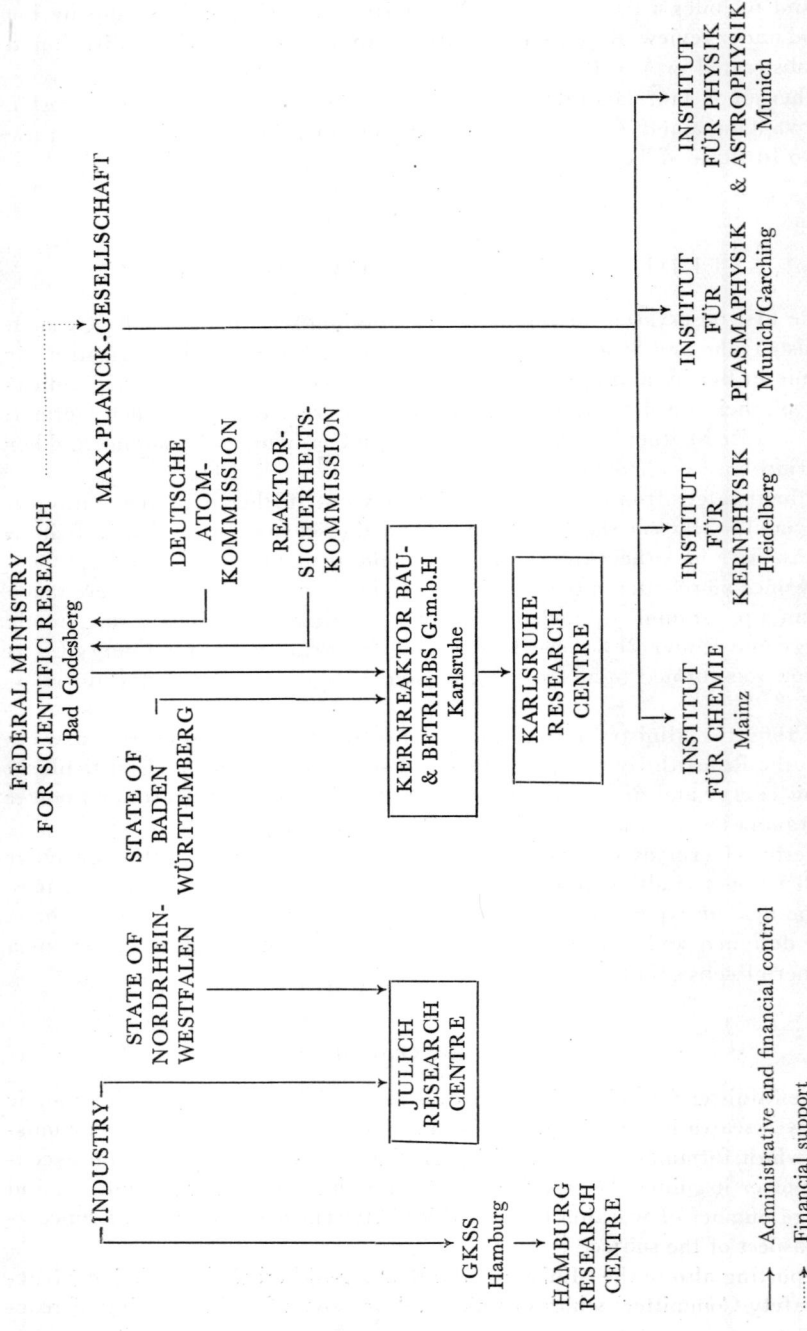

ATOMIC ENERGY ORGANIZATION OF WESTERN GERMANY

tors and other nuclear facilities and ensures that the Federal safety regulations are adhered to.

The Federal Government is not the only nuclear authority in Germany. Each of the States of Germany has its own nuclear programme and a Committee on Nuclear Affairs has been set up to co-ordinate these programmes with that of the central government. Many of the individual atomic energy projects are in fact jointly controlled by the Federal Government and the State Government and in many cases industry and the universities are also involved.

Official Nuclear Research Establishments

Kernforschungszentrum Karlsruhe

The Nuclear Research Centre at Karlsruhe is operated by Kernreaktor Bau- und Betriebs-Gesellschaft, m.b.H., a company now owned jointly by the Federal Government and the State of Baden-Württemberg. Karlsruhe is essentially a reactor centre where the first all-German reactors were designed and constructed. It also functions as the principal centre in Germany for training nuclear scientists and engineers.

The facilities of the Centre are described in a booklet [50] issued in 1963 and a detailed account of the research work of the individual laboratories of the Centre is given in a special issue [51] of *Atomwirtschaft* for April 1963.

The Centre was intended as a national laboratory but part of it is now being operated by Euratom as a section of the Common Research Establishment.

Kernforschungsanlage Jülich des Landes Nordrhein-Westfalen e.V. (KFA)

KFA is a company formed to build and operate the research centre at Jülich and to advise the State Government on atomic energy matters. The company is owned jointly by the Nordrhein-Westfalen State Government, the five scientific universities of the State and twelve industrial firms. Financial contributions to the Centre have also been made by the Federal Government.

The Jülich Research Centre is not yet completed and some of the constituent institutes are still located elsewhere. The research programme will cover high energy physics, fundamental studies, radiobiology, radiochemistry, reactor physics and technology, direct energy conversion and plasma physics.

Official Information Services

Because of its long experience in chemical documentation the Gmelin Institute for Inorganic Chemistry was chosen in 1957 by the Federal Minister for Atomic Energy to establish a national atomic energy documentation centre. Thus was created the Zentralstelle für Atomkernenergie Dokumentation which is the principal depository library in Germany for atomic energy publications and one of the main sources of information on German nuclear literature. All documents received by ZAED are indexed and listed in the *AED Index Bibliography,* which is published monthly, and described in greater detail in Part 4.

Reports are issued by the Karlsruhe Nuclear Research Centre and by the individual Institutes of the Research Centre at Jülich and these are widely distributed and listed in *N.S.A.* and in the AED bibliography. A complete list [52] of Karlsruhe reports and papers published from 1956 to 1962 has been issued by the Documentation Section of the Karlsruhe Library which also has a large collection of atomic energy reports and publications.

The Central Library at Jülich is now a depository library for atomic energy reports and provides a library and information service to all the institutes and outstations of KFA and to other organizations.

One of the largest collections of report literature in Germany is that of the Technische Informationsbibliothek, which is part of the Library of the Technische Hochschule at Hanover. The Informationsbibliothek has been established to collect and make available technical reports and, in this field, provides services similar to those of the National Central Library and National Lending Library in Britain.

OTHER SOURCES OF INFORMATION IN GERMANY

Max-Planck-Gesellschaft zur Förderung der Wissenschaften

Founded in 1911 as the Kaiser-Wilhelm Society, the Max Planck Society for the Promotion of Science has, for over fifty years, dominated scientific research in Germany. At the present moment forty-five separate Institutes are operated by the Society in Germany, although two of these, the newly created Institut für Dokumentationswesen at Frankfurt, and the Institut für Plasmaphysik at Munich are not officially listed as Max Planck institutes. In the nuclear field the most important of the institutes are the Institut für Chemie at Mainz, the Institut für Kernphysik at Heidelberg and the Institut für Physik and Astrophysik at Munich. In general the work of these institutes is reported in the scientific journals particularly *Zeitschrift für Naturforschung A & B,* which is published in co-operation with the Max-Planck institutes.

Hahn-Meitner-Institut für Kernforschung, Berlin

This is an autonomous institution financed largely by the Berlin authorities and consisting of four divisions devoted to nuclear chemistry, mathematics, radiation chemistry and nuclear physics respectively. The Institut issues annual reports and technical reports which are distributed to some foreign atomic energy projects and listed in the AED bibliography and in *N.S.A.* From time to time the Institut issues lists of its own publications which can be obtained from the Institut's library.

Gesellschaft für Kernenergieverwertung in Schiffbau und Schiffahrt m.b.H. (GKSS)

The Company for the Utilization of Nuclear Energy in Shipbuilding and Navigation is a consortia of thirty-six shipbuilding and engineering firms. It operates a research centre near Hamburg where design studies of nuclear reactors for ship propulsion are being carried out. One of the shareholders is Studiengesellschaft zur Förderung der Kernenergieverwertung in Schiffbau und Schiffahrt e.v. (Study Group for the Promotion of the Use of Nuclear Energy in Shipbuilding and Navigation), a scientific organization, of which over one hundred firms and government agencies are members. The Group has set up standing committees to deal with every aspect of ship propulsion and their work is described in the yearbook [53] published by the Group.

SWEDEN

Sweden has no oil and very little coal, but considerable water power resources which, it is estimated, will satisfy the country's energy requirements until the mid nineteen-seventies, after which it is planned that a significant fraction of the electricity required will be derived from nuclear sources and a first target of 2000 MW of nuclear power by 1975 has been set. This will enable Sweden to be largely self-sufficient from the energy point of view since she has large deposits of uranium-bearing oil shales. The Swedes are interested in nuclear power, not only from the point of view of electricity generation but as a large-scale source of heat for space heating, and the prototype power reactor recently completed at Agesta near Stockholm will be used mainly for district heating of a Stockholm suburb.

The Swedish Government is advised on all matters pertaining to atomic energy by an Atomic Energy Board (Delegationen för Atomenergifrågor), which is also responsible for licencing and for inspection and supervision of nuclear installations. Another body, the Atomic Research Council (Statens Råd för Atomforskning) advises the Government on fundamental nuclear research and allocates Government funds for this purpose to universities and institutes.

The responsibility for carrying out the Government's atomic energy programme is entrusted to Aktiebolaget Atomenergi (Atomic Energy Company of Sweden), a public company formed in 1947 to undertake research and development in all aspects of atomic energy and, in collaboration with the State Power Board, to implement the nuclear power programme.

Aktiebolaget Atomenergi

The Government holds four-sevenths of the shares in this company, the remainder being divided among private and municipal interests, and provides

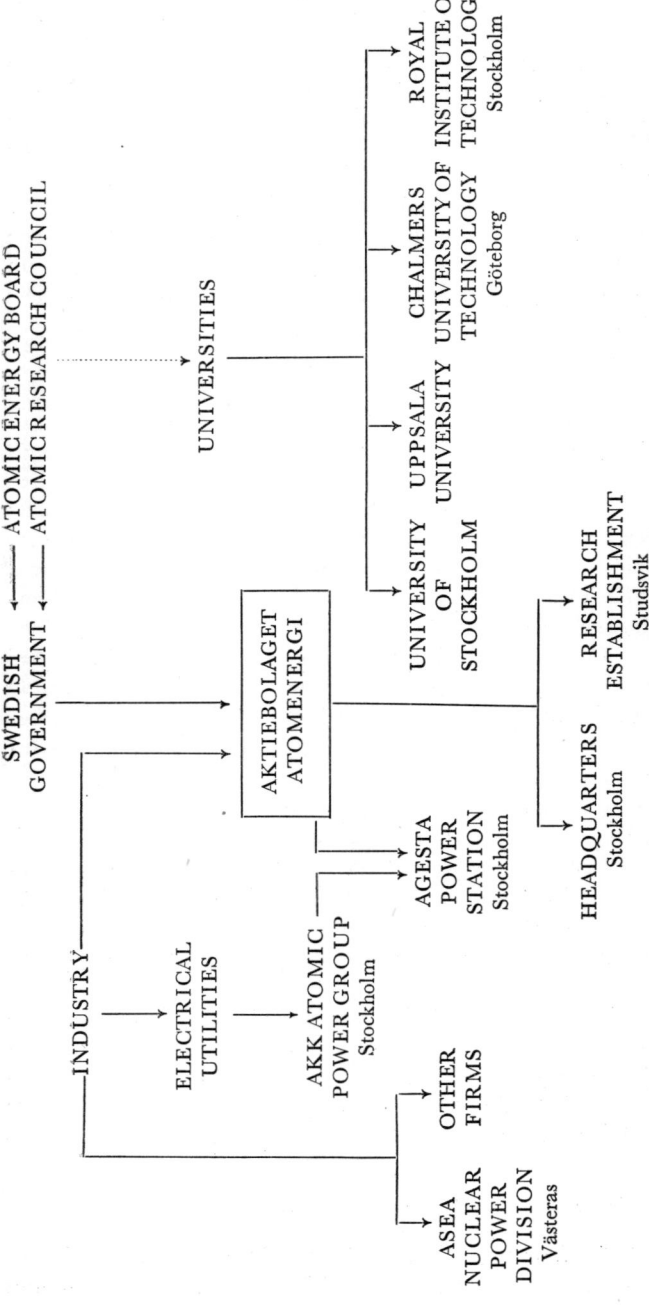

ATOMIC ENERGY ORGANIZATION IN SWEDEN

almost the whole of the money required for operating and for new capital expenditure. The headquarters and central laboratories are at Stockholm but the main research laboratory is at Studsvik, south of Stockholm.

Stockholm Laboratories

The first research reactor, R1, was built here in 1954 and is now used mainly for reactor physics experiments and the production of radioisotopes. Prospecting for uranium ores is controlled from here, the work being carried out in collaboration with the Geological Survey of Sweden. A uranium extraction plant and a fuel-element fabrication plant are also situated in Stockholm.

Studsvik Research Establishment, Nyköping

Construction of this laboratory began in 1955 and is now almost completed. The facilities include two research reactors, a high-flux materials testing reactor, a critical assembly, radiochemical laboratories, a metallurgy laboratory and the usual supporting services. The Swedish Atomic Research Council also operates a laboratory for fundamental studies at Studsvik.

A description of the research carried out by AB Atomenergi is given in a booklet, [54] issued in 1962, which describes the facilities at the various laboratories. More recent information is given in the annual reports [55] which are issued in both Swedish and English editions.

Information Services of AB Atomenergi

The technical reports published by AB Atomenergi are usually in English and are given a wide distribution in Sweden and externally. In Britain they are received by the UKAEA and the National Lending Library, and they are sent to the USAEC and abstracted in *N.S.A.* The library at Studsvik is the main depository for atomic energy publications and provides a service to other organizations in Sweden. The Library issues *Atomdokumentation* [56] in five series which, although intended mainly for internal use, is distributed to external organizations under certain circumstances. Newly issued reports are also listed in *Reaktorn,* [57] a house journal which contains news and review articles.

Other Sources of Information in Sweden

Universities

Fundamental research in the universities is sponsored by the Swedish Atomic Research Council, and the universities are encouraged to make use of facilities available at the Studsvik Research Establishment and the AB Atomenergi Laboratories in Stockholm. The universities chiefly concerned are those of

Stockholm and Uppsala together with the Royal Institute of Technology in Stockholm and Chalmers University of Technology, Göteborg. In most cases the results of this research are published in the scientific and technical journals.

Industry

A nuclear consortia, Atomkraftkonsortiet Krangede, A.B. (AKK Atomic Power Group), of which the constituent companies are private and municipal electrical utilities, has been set up to operate the Agesta power station and to plan further power stations.

The largest single industrial organization in the nuclear field is Allmanna Svenska Electriska A.B. (ASEA), a vast electrical manufacturing organization with a nuclear power department which is equipped to design and construct complete nuclear power stations and their components. ASEA is the main contractor for the Agesta prototype power reactor and has operated its own atomic energy research laboratories since 1955. ASEA issues a large number of internal reports which are given only a limited distribution, but many of these are written up for publication and appear either in scientific journals or in *ASEA Research* [58] which is published in English. The company also issues a journal [59] which covers all aspects of its work and contains technical articles on its nuclear activities.

BELGIUM

The uranium mines of the Belgian Congo were the major sources of uranium for the war-time atomic bomb project in American, and the bilateral agreements under which uranium ores were provided to the U.S.A. and the U.K. gave Belgium direct access to a certain amount of information on the processing of fissile materials and the technology of nuclear reactors, so that its was possible to undertake a limited nuclear programme at a comparatively early stage.

Atomic Energy Organization

A Commissariat à l'Énergie Atomique was set up in 1951 to co-ordinate nuclear activities in Belgium, to promote research and development in this field and to advise the Government on the policies which it should adopt. At the same time a Nuclear Applications Department was created in the Ministry of Economic Affairs to co-ordinate work on the industrial applications of nuclear energy.

At the operating level, research and development has been made the responsibility of the Centre d'Étude de l'Énergie Nucléaire, a public utility sponsored jointly by the Government, industry and the universities. Nuclear research in the universities is co-ordinated by the Institut Interuniversitaire des Sciences Nucléaires which is sponsored by the Fonds National de la Recherche Scientifique.

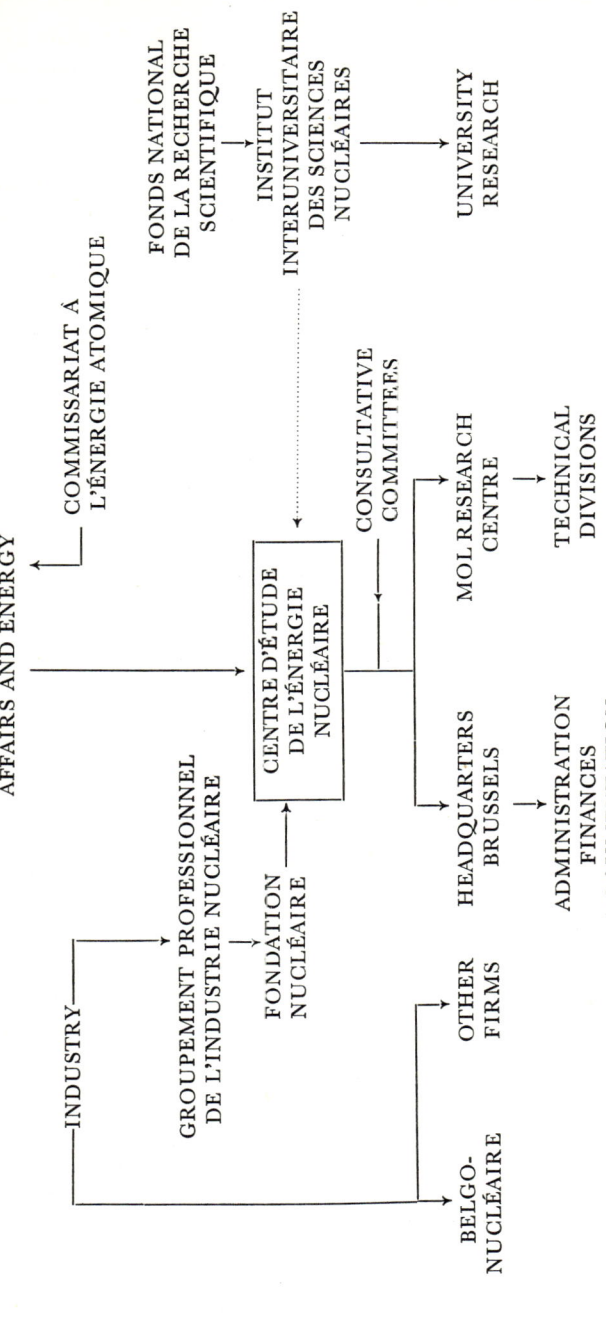

ATOMIC ENERGY ORGANIZATION IN BELGIUM

Centre d'Étude de l'Énergie Nucleaire (CEN)

Founded in 1952, with headquarters in Brussels, CEN is the principal nuclear research organization in Belgium. It operates a research centre at Mol, about 30 miles east of Antwerp, where the facilities include two research reactors, a prototype power reactor and associated laboratories. A plutonium-processing laboratory, which is operated jointly by CEN and Belgonucléaire has also been built on the site, and not far away is the fuel processing plant being built by Eurochemic, a company established by the European Nuclear Energy Agency of the Organization for Economic Co-operation and Development. The work of the Mol Research Centre is mainly concerned with investigations in reactor technology and reactor materials, although there are laboratories for medical and biological research, and an important function of the Centre is the training of nuclear technologists for subsequent employment in industry.

A descriptive booklet [60] on the Mol laboratory was issued in 1959 and more recent information can be found in the annual reports, [61] each of which includes a list of conference papers, journal articles and reports published by CEN staff during the period under review. CEN reports are sent to the UKAEA, the USAEC and other foreign projects and are abstracted in *N.S.A.* The Documentation Centre in Brussels is the main Belgian depository for atomic energy publications and provides a service to other Belgian organizations. Since January 1964 the Centre has issued a weekly *Bulletin d'Information Technique* which is both an information bulletin and an accessions list of material received.

Other Sources of Information in Belgium

Industry has played an important part in the development of atomic energy in Belgium, not only by the formation of groups and companies to manufacture and supply nuclear plant and equipment, but also by stimulating nuclear research and promoting the establishment of research laboratories. The Groupement Professionnel de l'Industrie Nucléaire, established in 1957 to represent industry in nuclear matters, arranges the financial contribution which industry makes for the support of CEN, by a levy on its member firms. The Association Belge pour le Développement Pacifique de l'Énergie Atomique (BELGICATOM) is the Belgian equivalent of Atomic Industrial Forum, although its main function is to inform industry and the general public on atomic energy developments. To this end it publishes a bimonthly journal [62] which contains technical papers by CEN staff and notes of nuclear developments in Europe and elsewhere, and lists the current contents of the world's nuclear engineering journals.

A nuclear consortia, the Société Belge pour l'Industrie Nucléaire (BELGONUCLEAIRE), has been formed whose activities lie mainly in the fields of reactor design and construction and the manufacture of plutonium-enriched fuels. BELGONUCLEAIRE is also developing the VULCAIN propulsion reactor in collaboration with the UKAEA and has contracts to build two experimental fast reactors at the French CEA establishment at Cadarache. More information on the work of this organization can be found in the annual reports [63] which are published in English and French.

The largest single firm operating in the nuclear field is Ateliers de Constructions Electriques de Charleroi (ACEC) which is organized to build complete reactor installations or components, particle accelerators and nuclear radiation detectors. The Company publishes a quarterly journal, *ACEC Review*, [64] which contains articles by ACEC staff on nuclear research in the company.

Altogether there are a great many organizations involved in nuclear activities in Belgium, particularly in the industrial sector, and a comprehensive guide [65] to these has been issued by the Groupment Professionnel de l'Industrie Nucléaire. A more detailed description of the research work of the various official, industrial and educational organizations is contained in a report issued by CEN [66] in 1959.

NETHERLANDS

Atomic Energy Organization

A Nuclear Energy Committee was set up in 1955 to develop an atomic energy policy and to advise the Government on atomic energy matters. This is essentially a consultative committee, representing the various ministries of the Netherlands Government, and operating through the Directorate of Nuclear Energy which is a division of the Ministry of Economic Affairs. As a result of recent legislation, a Central Council for Nuclear Energy has been established, with two sub-councils, the Industrial Council for Nuclear Energy and the Scientific Council for Nuclear Energy, to advise the Government and individual ministers, as well as other official bodies, and to co-ordinate national and international activities in the atomic energy field.

The executive responsibility for nuclear research and development lies with the Reactor Centrum Nederland (RCN), an organization founded in 1955 with government and industrial support. RCN developed, in the first place, from the Reactor Committee of the Foundation for Fundamental Research on Matter (FOM), a government sponsored organization established in 1946 to co-ordinate basic research in the physical sciences. In 1951 FOM signed an agreement with the Norwegian Institutt for Atomenergi to co-operate in the design and construction of a research reactor at the Joint Establishment for Nuclear Energy Research (JENER) at Kjeller in Norway, and since then has taken a leading part in the development of atomic energy in Holland.

Although the Netherlands has little in the way of natural resources and has to rely on imported coal for a substantial fraction of her electrical energy, the approach to nuclear power has been wary, to say the least, and the construction of the first power station is unlikely to begin before 1965. However

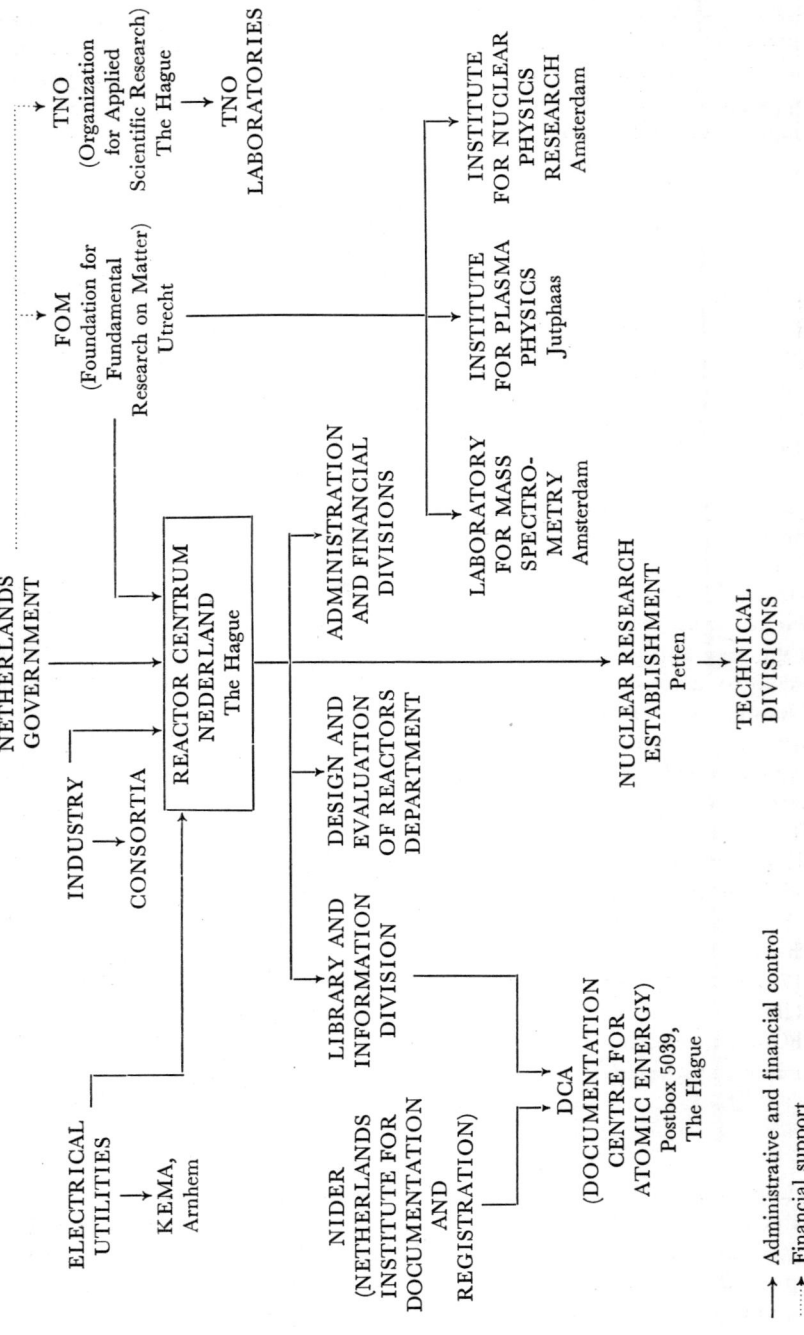

ATOMIC ENERGY ORGANIZATION IN THE NETHERLANDS

industry in Holland has been extremely active in developing the skills and experience necessary for a nuclear power programme and a good deal of research is being undertaken. A fairly complete picture of the present state of development can be obtained from the special issue of *Atomwirtschaft* [67] for November 1963, which is devoted entirely to atomic energy in the Netherlands.

Reactor Centrum Nederland

The Netherlands Government, the electrical utility companies, FOM, and Dutch industry are all represented on the Board of Directors of RCN, although the larger part of its financial resources come from Government funds. The main object of the organization is to obtain scientific and technical experience in the atomic energy field and to make the knowledge thus gained generally available to Dutch institutions.

RCN operates a research centre at Petten, carries out research at the Norwegian centre at Kjeller, is co-operating with the electrical utility companies in the development of an aqueous, homogeneous suspension reactor, and with FOM in a project to separate uranium isotopes by means of a high speed centrifuge. It also has contracts with a number of universities to carry out specific projects.

At the Nuclear Research Establishment at Petten, two reactors and a critical assembly are now in operation, although one of these, a high-flux reactor, has been handed over to Euratom as part of an agreement which allows Euratom to build a section of its Common Research Centre on the Petten site. The laboratory is engaged mainly in reactor research, one of the projects being a feasibility study of a propulsion reactor to which Euratom is contributing.

A description of the work of RCN will be found in the November 1962 issue of RCN's own journal, *Atoomenergie,* [68] a special issue devoted entirely to the work of the Petten centre. More information on the main facilities is given in a booklet [69] issued in 1963. RCN also publishes an annual report in Dutch and, from time to time, a summary of activities in English, the last one [70] covering the period January 1962 to June 1963. The headquarters of RCN in the Hague houses the secretariat of the Netherlands Atom Forum which publishes conference proceedings, a monthly news letter, and various other items of information.

RCN Information Service

The library of RCN is the main depository collection of atomic energy literature in Holland and provides a service to other organizations and research laboratories. A Documentation Centre for Nuclear Energy (DCA) was established in 1958 in co-operation with the Netherlands Institute for Documentation and Registration (NIDER) and is prepared, subject to certain conditions, to carry out information searches at the request of individuals or organizations, using the facilities of the libraries at RCN and NIDER. A subscription service is also available by which subscribers can receive, each month, selections of annotated references to current literature in the fields of reactor engineering, nuclear physics and nuclear chemistry. A small number of reports is issued by RCN and these are sent to Euratom countries, the UKAEA and the USAEC and abstracted in *N.S.A.* RCN also publishes a monthly journal [71] which is given a wide circulation and which contains original articles on the research carried out.

Much of the early Dutch research work was done at JENER, in Norway, and the results published as JENER reports which appear in *N.S.A.* and were sent to Britain, Canada and the U.S.A. and other foreign projects.

FOUNDATION FOR FUNDAMENTAL RESEARCH ON MATTER

Stitchting voor Fundamenteel Onderzoek der Materie (FOM) is a government sponsored organization operating a number of laboratories of which the most important in the nuclear field are:

Laboratory for Mass Spectrometry, Amsterdam

Studies of centrifugal methods of isotope separation are being undertaken in collaboration with RCN.

FOM Institute for Plasma Physics, Jutphaas

This is the main centre for fusion research in Holland, although some plasma physics work is being done at other laboratories. Part of the programme at Jutphaas is supported by Euratom.

Institute for Nuclear Physics Research, Amsterdam

A laboratory established jointly by FOM, the Amsterdam municipality, and Philips Industries, to carry out research into applications of radioisotopes. It also operates an Isotope School.

In general, results of research are published in scientific journals but a few reports have been issued and abstracted in *N.S.A.* FOM issues an annual report in Dutch, the 1959 issue [72] of which contained a complete review of FOM's activities between 1946 and 1960 (with a summary in English) and details of the four year research programme for 1961–1964.

OTHER SOURCES OF INFORMATION IN THE NETHERLANDS

For a small country with only a moderate nuclear programme Holland has an unusual multiplicity of groups and organizations active in this field. Only a few of these can be mentioned but more information can be found in a guide [73] to these organizations compiled by the Nuclear Energy Committee in 1960.

Netherlands Organization for Applied Scientific Research (TNO)

This is an organization with functions rather like those of the British Department of Scientific and Industrial Research, with a number of laboratories, many of which are working on nuclear problems. The most important of these are the Central Technical Institute, The Hague, which is studying heat transfer and reactor coolants, the Central Laboratory, Delft which is working on the uses of radioisotopes, the Nuclear Propulsion Department of the Research Centre for Building and Navigation, Amsterdam, which studies problems connected with the use of reactors as marine propulsion units and houses the main Dutch documentation centre in this field, and the Institute for the Application of Nuclear Energy to Agriculture which has entered into an agreement with Euratom for research into the use of ionizing radiations in plant genetics and for food preservation. TNO issues an annual report and a news journal [74] which describes its major activities in atomic energy and other fields.

Tot Keuring van Electrotechnische Materialien NV (KEMA)

Testing of Electrical Materials Company is a subsidiary of Samenwerkende Electriciteits Productiebedrijwen (Co-operating Electricity Producers), a co-operative association of nine electrical utilities which is represented on the Board of Governors of RCN. KEMA operates laboratories at Arnheim, where research on the suspension reactor is being done in collaboration with RCN, and issues scientific and technical reports which are distributed to a number of foreign official projects and abstracted in *N.S.A.*

Industry

The largest industrial consortia is NERATOOM N.V., an association of nine companies which, apart from building nuclear plant in Holland, has successfully completed two contracts for reactor installations in Italy. A description of the work of NERATOOM and of other firms in the nuclear field is given in a special number of *Atoomenergie* [75] published in 1961.

SWITZERLAND

Switzerland has no coal or oil but ample hydroelectric power so that there is no immediate stimulus to launch a full-scale nuclear power programme. However, since there is likely to be a need for nuclear power in the not too distant future, the groundwork of an atomic energy industry is being laid; four research reactors are now being operated and a prototype power reactor is under construction.

Atomic Energy Organization

The initiative for developing an atomic energy programme in Switzerland came mainly from industry, particularly the large organizations such as Brown

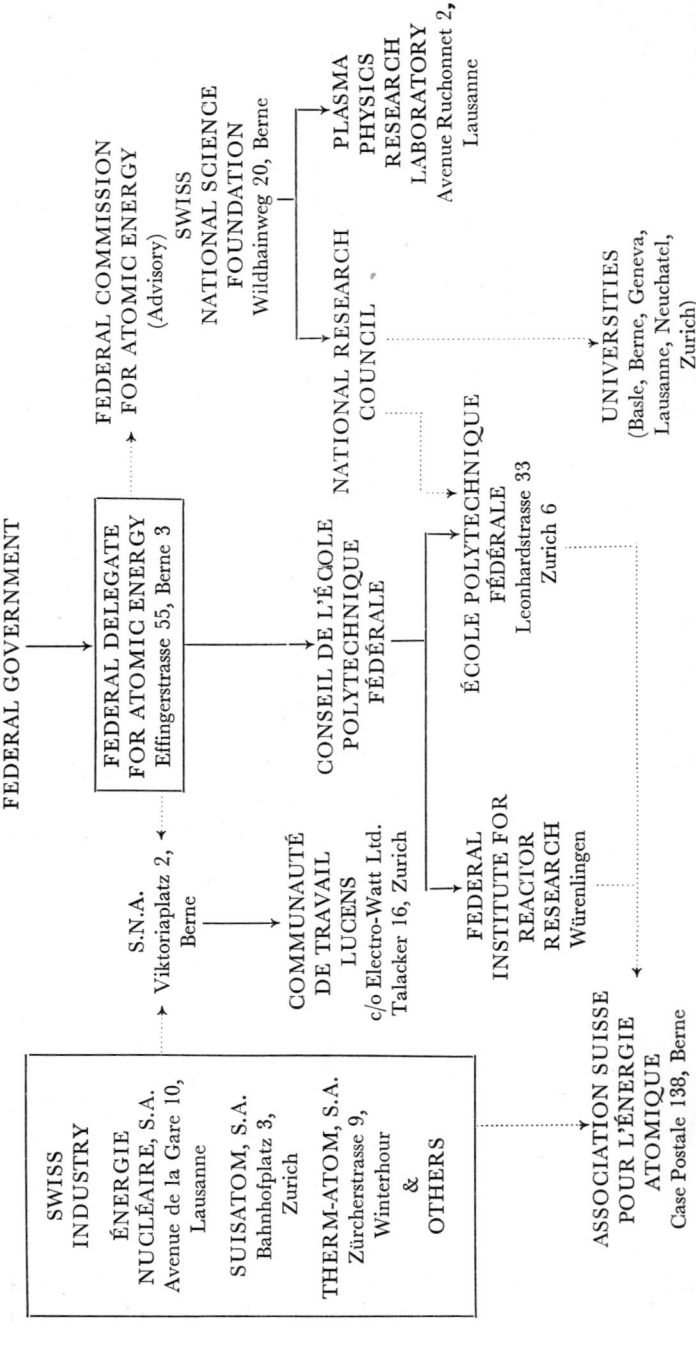

ATOMIC ENERGY ORGANIZATION IN SWITZERLAND

Boveri, and in 1955 a private corporation, Reaktor AG, was formed with the general participation of Swiss industry and with some financial support from the Federal Government. It established a research laboratory at Würenlingen, where Switzerland's first two research reactors were built, and issued a series of scientific and technical reports on the work of the laboratories.

In 1958 a Federal Atomic Energy Commission was set up as an advisory body under the chairmanship of the Federal Delegate for Atomic Energy, and the office of the Federal Delegate became the official organization for dealing with atomic energy questions in Switzerland. The Delegate answers to the Minister of Transport, Communications and Energy and is responsible for the encouragement and co-ordination of atomic energy research and development, for drafting appropriate legislation and ensuring that it is applied and for collaboration between the Government and private industry. The Delegate also represents the Government on international atomic energy organizations.

In 1962 the assets of Reaktor AG were taken over by the Federal Government and the Würenlingen laboratories were attached to the École Polytechnique Fédérale, Zurich and renamed the Federal Institute for Reactor Research (Eidgenossisches Institut für Reaktorforschung). The Institute is administered by the Board of Governors of the École Polytechnique who are advised by a Consultative Committee of which the Federal Delegate is a member. A detailed account of the work of the Institute and its facilities is given in a special issue [76] of *Neue Technik* published to commemorate the starting up of the reactor DIORIT in August 1960.

Apart from Würenlingen, the only other laboratory devoted entirely to atomic energy is the Plasma Physics Research Laboratory at Lausanne, which is engaged in nuclear fusion research. This laboratory was established in 1961 by the Swiss National Science Foundation. A good deal of nuclear research, is, however, carried out in the laboratories of the École Polytechnique Fédérale at Zurich, the École Polytechnique of the University of Lausanne, and in the Universities of Basle and Geneva, both of which have installed research reactors.

In the industrial field a company known as Société Nationale pour l'Encouragement de la Technique Atomique Industrielle (SNA) has been formed by three Swiss consortia, Énergie Nucléaire S.A., Lausanne; Suisatom S.A., Zurich; and Therm-Atom S.A., Zurich; to build a prototype gas-cooled, heavy-water-moderated power reactor at Lucens. Fifty per cent of the stock of SNA is held by the Federal Government, the other 50 per cent being owned by the consortia and other Swiss firms.

Information Services

There are small documentation sections in the research establishments at Würenlingen and Lausanne and a number of reports have been issued. Most of these are distributed to the major foreign atomic energy projects and are abstracted in *N.S.A*. The main depository library for atomic energy documents in Switzerland is, however, the library of the École Polytechnique Fédérale at Zurich. The library has a Documentation Centre which issues reports and also provides a bibliographical service of abstracts cards [77] which are issued in three series and can be purchased on subscription.

Another important source of information is the Swiss Nuclear Society (Schweizerische Gesellschaft von Fachleuten der Kerntechnik), with headquarters at Würenlingen, which is an association of nuclear engineers, scientists, technicians and others working in the atomic energy field. In co-operation with the Swiss Federation for Automatic Control it publishes *Neue Technik* [78] which carries original articles and news items on nuclear engineering and automation and is chiefly interesting for the very detailed accounts of the atomic energy organization of the major western European countries which it publishes from time to time.

As in nearly all European countries, Switzerland has an atomic energy forum, the Association Suisse pour l'Énergie Atomique (ASPEA) with offices in Berne. The aims of ASPEA are similar to those of Atomic Industrial Forum and it has been active in arranging meetings and conferences and in ensuring that Swiss industry and the public is well informed on events in Switzerland and in other countries. ASPEA publishes a *Bulletin* [79] which carries short news articles and longer surveys of nuclear developments, but its most useful publication is a manual [80] which covers every aspect of atomic energy in Switzerland, including all Swiss legislation and regulations relating to atomic energy.

NORWAY

Norway has ample water power sources and her interest in nuclear power lies not so much in the field of electricity generation, as in process heat and marine propulsion applications, as well as a desire to ensure that Norwegian firms can compete in the world's atomic energy industry.

A decision to build a heavy-water-moderated reactor was taken in 1947 and an Institutt for Atomenergi was established to carry out the necessary development work. The uranium was supplied by the Netherlands Government and an agreement for co-operation between the two countries resulted in the creation of the Joint Establishment for Nuclear Energy Research (JENER), at Kjeller where the first reactor was built. This agreement lasted until 1959, when the Petten Research Establishment came into operation in Holland, and a new agreement was concluded whereby the two countries developed their own research programmes but continued to carry out specific joint projects at both Kjeller and Petten.

Atomic Energy Organization

The Institutt for Atomenergi (IFA) is an independent foundation, financed mainly by the Government through the Ministry for Industry, and having a governing body of six members, two each being appointed by the Norwegian Government, Norsk Hydro and the Royal Norwegian Council for Scientific Research. The Kjeller Research Laboratory now has three research reactors in operation and design studies on a marine boiling water reactor are being undertaken. IFA also operates the establishment at Halden, 80 miles south of Oslo, which is the site of the European Nuclear Energy Agency heavy-water, boiling water reactor project.

Information Services

The research work carried out at Kjeller from its foundation until 1959 is described in the JENER annual reports [81] which also contain lists of the technical reports and journal articles published during the period. More recently the work of IFA has been described in a booklet [82] issued in 1961 and in the activities report, the last issue [83] of which covers the period up to 1960.

These reports and booklets are issued by the Technical Information Service of IFA which is based on the Library at Kjeller. This is the main depository collection of atomic energy publications in Norway and provides a service to other Norwegian organizations. The Information Service maintains an extensive photographic collection and has produced a number of films some of which are available in both Norwegian and English versions.

Most of the JENER reports were published in English, given a very wide distribution and abstracted in *N.S.A.* The reports now issued by IFA are mainly internal although a few are distributed to foreign atomic energy projects. Research results are made known in articles written for the major scientific and technical journals, particularly those in the English language, although a good many technical papers are published in the *Technical Weekly* [84] of the Norwegian Engineer's Association.

DENMARK

The birthplace of Niels Bohr has long been a leading contributor to knowledge in the field of nuclear physics, mainly through the work of the Institute for Theoretical Physics of Copenhagen University of which Bohr was Director before the Second World War. Later, in 1955, when an Atomic Energy Commission was formed, Bohr became chairman and remained so until his death in 1962.

Denmark has little in the way of energy resources and imports 90 per cent of her fuel requirements, and although a full-scale nuclear power programme has not yet evolved, design studies have been made of reactors for electrical generation and marine propulsion.

Danish Atomic Energy Commission

The Commission has its headquarters in Copenhagen although the principal research laboratory is at Risø, twenty miles west of the capital. The Risø establishment now has three research reactors and a linear accelerator and studies cover all aspects of atomic energy. A particular feature is the number of irradiation experiments being carried out for foreign projects, including the UKAEA, the French CEA, AB Atomenergi of Sweden, and the OECD Dragon project. The annual report [85] of the Commission is published in English and Danish and gives a detailed account of the research being done as well as a list of reports and journal articles issued during the year under review.

The library at Risø is the main depository library in Denmark for atomic energy documents and provides a service to other Danish organizations. Risø reports are issued through the library and are distributed to all the major foreign projects and abstracted in *N.S.A.*

Danatom

DANATOM, the Danish Association for Industrial Development of Atomic Energy was founded in 1956 as an institution of the Danish Academy of Technical Sciences. Its members include private firms, electrical utilities and the Atomic Energy Commission and its main functions are to assist Danish enterprises in matters relating to the design and construction of nuclear power plants and to secure Danish participation in foreign nuclear projects. At present DANATOM is carrying out design studies for a ship propulsion reactor, a gas-cooled power reactor and a boiling water power reactor. A small number of technical reports are issued, mostly in English, although they are not given a wide distribution. They are listed each year in the annual reports [86] which are also published in English.

REFERENCES

France

1. COMMISSARIAT A L'ÉNERGIE ATOMIQUE. *The Saclay nuclear research centre*. (In English.) CEA, Service des Relations Publiques, 15 rue du Capitaine Scott, Paris 15e. 1962.
2. COMMISSARIAT A L'ÉNERGIE ATOMIQUE. *The Fontenay-aux-Roses nuclear research centre*. (In English.) CEA, Service des Relations Publiques, Paris. 1961.
3. COMMISSARIAT A L'ÉNERGIE ATOMIQUE. *The Grenoble nuclear research centre*. (In English.) CEA, Service des Relations, Paris. 1962.
4. COMMISSARIAT A L'ÉNERGIE ATOMIQUE. *The Cadarache nuclear research centre*. (In English.) CEA, Service des Relations Publiques, Paris 1963.
5. COMMISSARIAT A L'ÉNERGIE ATOMIQUE. *The French Atomic Energy Commission, 1945—1960*. (In English.) CEA, Service des Relations Publiques, Paris. 1960.
6. COMMISSARIAT A L'ÉNERGIE ATOMIQUE. *Rapport annuel.* 1945 and onwards. CEA, 69 rue de Varenne, Paris 7e.

7. COMMISSARIAT A L'ÉNERGIE ATOMIQUE. *Developments and programmes.* (In English.) CEA, Service des Relations Publiques, Paris.
8. COMMISSARIAT A L'ÉNERGIE ATOMIQUE. *Liste récapitulative des rapports CEA publiés par le Commissariat à l'Énergie Atomique 1948–1961* (with supplements). Service de Documentation du CEA, Centre d'Études Nucléaires de Saclay, Boite Postale No. 2, Gif-sur-Yvette (Seine et Oise), France. 1962.
9. C.E.A. CENTRE D'ÉTUDES NUCLÉAIRES DE GRENOBLE. *Publications et notes scientifiques et techniques.* Group de Documentation, Centre d'Études Nucléaires de Grenoble, Boite Postale No. 269, Grenoble, France. 1963.
10. *Bulletin d'Informations Scientifiques et Techniques.* (Monthly.) Éditions Dunod, 92 rue Bonaparte, Paris 6e. 1957 and onwards.
In French with titles and abstracts in English.
11. *Notes d'Information.* (Irregular.) CEA, Service des Relations Publiques, Paris.
12. *Physinedx.* (Irregular.) *Series A. Plasma physics and controlled fusion. Series B. Solid state physics. Series C. High energy physics. Series D. Nuclear physics.* Service de Documentation du CEA, Centre d'Études Nucléaires de Saclay, Boite Postale No. 2, Gif-sur-Yvette (Seine et Oise), France. 1963 and onwards. Each issue comprises (1) a bibliography of current reports and articles; (2) a KWIC subject index; (3) an author index; and (4) a report number index. All entries are in English and a list of the journals scanned is given in each issue.
13. COMMISSARIAT A L'ÉNERGIE ATOMIQUE. *La documentation scientifique et technique au CEA.* Service de Documentation du CEA, Centre d'Études Nucléaires de Saclay, Boite Postale No. 2, Gif-sur-Yvette (Seine et Oise), France. 1963.
14. COMMISSARIAT A L'ÉNERGIE ATOMIQUE. *Cinemathèque: A-Films tous publics; B-Films pour publics spécialises.* CEA, Service des Relations Publiques, Paris. 1963.
15. ÉLECTRICITÉ DE FRANCE. *La centrale nucléaire de Chinon.* EDF, Region d'Équipement Thermique Nucléaire No. 1, 2 avenue de la Libération, Clamart (Seine), France. 1961.
16. ÉLECTRICITÉ DE FRANCE. *Rapport d'activité: comptes de gestion.* EDF, 3 rue de Messine, Paris 8e.
17. *Bulletin d'Information A.T.E.N.* (Bimonthly.) Association Technique pour l'Énergie Nucléaire, 26 rue de Clichy, Paris 9e. 1956 and onwards.
18. *Courrier de l'A.T.E.N.* (Semi-monthly.) Association Technique pour l'Énergie Nucléaire, 26 rue de Clichy, Paris 9e. 1963 and onwards.
19. *Propriété Industrielle Nucléaire.* (Bimonthly.) BEVATOME, 25 rue de Ponthieu, Paris 8e. 1958 and onwards.
Summaries in French and English.
20. *Annuaire de l'Activité Nucléaire Française.* Groupement Intersyndical de l'Industrie Nucléaire, 64 avenue Marceau, Paris 8e *and* Association Technique pour l'Énergie Nucléaire, 4 rue de Teheran, Paris 8e. 1961.
Three language edition in French, English and Italian.

Italy

21. COMITATO NAZIONALE PER LE RICHERCHE NUCLEARI. *A five year plan for development of nuclear research in Italy* (In English and Italian editions.) CNRN, via Belisario 15, Rome. 1958.
22. COMITATO NAZIONALE PER LA RICHERCHE NUCLEARI. *Activities report 1958 to 1959.* (In English and Italian editions.) CNRN, Rome. 1960.
23. COMITATO NAZIONALE PER I'ENERGIA NUCLEARI. *Rapporto di attivita.* CNEN, via Belisario 15, Rome. 1961 and onwards.
24. COMITATO NAZIONALE PER LE RICHERCHE NUCLEARI. *C.N.R.N., 1952–1959.* (In English.) Publications Office, CNRN, Rome. 1959.

25. COMITATO NAZIONALE PER LE RICHERCHE NUCLEARI. *The Centre for Nuclear Studies, Ispra.* (In English and Italian editions.) Publications Office, CNRN, Rome. 1959.
26. COMITATO NAZIONALE PER LE RICHERCHE NUCLEARI. *The National Synchrotron Laboratory, Frascati.* Publications Office, CNRN, Rome. 1959.
27. COMITATO NAZIONALE PER L'ENERGIA NUCLEARI. *Il Centro della Casaccia.* Publications Office, CNEN, Rome. 1962.
28. COMITATO NAZIONALE PER L'ÉNERGIA NUCLEARI. *Rapporti tecnici: catalogo 1958—1961.* Publications Office, CNEN, Rome. 1962.
    Reports and articles are listed in 18 series under CN, CNI and CNEN numbers, with author index.
29. COMITATO NAZIONALE ENERGIA NUCLEARE. *Rapporti tecnici: catalogo 1962 and onwards.* Publications Office, CNEN, Rome. 1963 etc. (Annual).
    From 1962 reports are listed in 14 series under RT (technical reports) and RTI (internal technical reports) numbers with author indexes.
30. CNEN. LABORATORI NAZIONALI DI FRASCATI. *List of report and abstracts 1953 to 1961.* Servizio Documentazione del Laboratori Nazionale di Frascati del CNEN, Casella Postale 70, Frascati, Rome. 1962. (LNF 62/21).
31. CNEN. LABORATORI NAZIONALI DI FRASCATI. *Activity at the National Laboratories of Frascati.* (In English and Italian editions.) Servizio Documentazione, etc., Frascati, Rome. 1956 and onwards (half-yearly).
32. *Notiziario del CNEN.* (Monthly.) CNEN, via Belisario 15, Rome. 1955 and onwards.
    Surveys nuclear developments in Italy and abroad with emphasis on Italian and foreign regulations relating to nuclear energy, and reports on the economic, legal and health and safety aspects of the subject.
33. CENTRO INFORMAZIONI STUDI ESPERIENZE. *CISE Laboratories.* (In English.) Documentation Service, CISE, Casella Postale 3986, Milan. 1959.
34. CENTRO INFORMAZIONI STUDI ESPERIENZE. *List of CISE's publications.* (With annual supplements.) Documentation Service, CISE, Casella Postale 3986, Milan. 1960.
35. *Atomo e Industria.* (Semi-monthly.) Edizioni " Atomo e Industria ", via Paisiello 26/28, Rome. 1957 and onwards.
    Articles and news items in English and Italian.
36. *Rivista di Ingegneria Nucleare.* (Bimonthly.) Associazone Nazionale d'Ingegneria Nucleare, Piazza Sallustio 24, Rome. January 1963 and onwards.

Japan

37. JAPAN ATOMIC ENERGY COMMISSION. *Atomic Energy in Japan.* (In English.) Atomic Energy Bureau, 3 Kasumigaseki, Chiyoda-ku, Tokyo. 1962.
38. JAPAN ATOMIC ENERGY COMMISSION. *Long-range programme on development and utilization of atomic energy.* (In English.) Atomic Energy Bureau, Chiyoda-ku, Tokyo. 1961.
39. JAPAN ATOMIC ENERGY COMMISSION. *Japan Atomic Energy Research Institute: a brief guide.* (In English.) J.A.E.R.I, 1—1 Tamura-cho, Minato-ku, Tokyo. 1963.
40. *Nuclear Science Abstracts from Japan.* (Irregular.) Division of Techncial Information, Tokai Research Establishment, Tokai-mura, Naka-gun, Ibari-ken, Japan. 1961 and onwards. (In English.)
    English translations of many of the Japanese articles abstracted may be obtained by direct application to the Tokai Division of Technical Information.
41. *Gen-ken (Nuclear research.* (Monthly.) J.A.E.R.I., Minato-ku, Tokyo. 1960 and onwards.

42. *Atoms in Japan.* (Monthly.) Japan Atomic Industrial Forum, 1—1 Tamura-cho, Minato-ku, 1957 and onwards. (In English.)
43. *Genshiryoku Nenken (Atomic Energy Yearbook).* Japan Atomic Industrial Forum, Minato-ku, Tokyo. 1957 and onwards.
44. *JAIF's directory to Governmental agencies, associations and federations, industrial firms, universities, institutions and local governments in the field of atomic energy in Japan. 1958—1959.* (In English.) Japan Atomic Industrial Forum, Minato-ku, Tokyo. 1959.
45. JAPAN ATOMIC POWER CO. *Report 1960.* (In English.) J.A.P.Co., Otemachi 1-chome, Chiyoda-ku, Tokyo. 1961.
46. *Atomic industry of Japan (1963 special issue of " Atoms in Japan").* Japan Atomic Industrial Forum, Minato-ku, Tokyo. 1963.
   Includes a survey of the development of the atomic energy industry. (In English.)
47. *Journal of the Atomic Energy Society of Japan.* (Monthly.) Atomic Energy Society of Japan, c/o J.A.E.R.I., Tokyo. 1959 and onwards.
   Contents list and main articles in English; other articles in Japanese with English summaries. Each issue contains abstracts in Japanese of important articles and reports from world nuclear literature.
48. TOKYO UNIVERSITY. INSTITUTE OF NUCLEAR STUDY. *Annual report (latest issue covers the two years April 1960—March 1962).* Institute of Nuclear Study, Tanashi-machi, Kitatama-gun, Tokyo. 1955 and onwards.

Western Germany

49. HAHN, O. AND STRASSMANN, F. *Über den Nachweis und das Verhalten der bei der Bestrahlung des Urans mittels Neutronen entstehenden Erdalkalimetalle.* Naturwissenschaften, vol. 26, pp. 756. 1938.
50. *Kernforschungszentrum Karlsruhe.* (In German.) Kernreaktor Bau- und Betriebs-Gesellschaft m.b.H. Weberstraße 5, Karlsruhe. 1963.
51. *Kernforschungszentrum Karlsruhe.* (In German with English summaries.) *In:* Die Atomwirtschaft, vol. 8, no. 4, April 1963.
52. KERNFORSCHUNGSZENTRUM KARLSRUHE. *Liste der wissenschaftlichen Veröffentlichungen des Kernforschungszentrums Karlsruhe aus den Jahren 1956—1962.* Kernreaktor Bau- und Betriebs-Gesellschaft m.b.H., Karlsruhe. 1963. (KFK 160).
53. *Jahrbuch der Studiengesellschaft zur Förderung der Kernenergieverwertung in Schiffbau und Schiffahrt, e.v.* 34 Neuer Wall, Hamburg 36. 1957 and onwards.

Sweden

54. *Aktiebolaget Atomenergi: the Atomic Energy Company of Sweden.* (In English.) Information Office, AB Atomenergi, 5—7 Lövholmsvägen, Stockholm 9. 1962.
55. AKTIEBOLAGET ATOMENERGI. *Annual report.* (In English.) Information Office, AB Atomenergi, Stockholm 9. 1959 and onwards.
56. *Atomdokumentation.* AB Atomenergi, Biblioteket, Studsvik, Nyköping, Sweden. Published in five series as follows:

   1. Boklista
   2. Mikrokortlista
   3. Rapportlista
   } These are accessions lists compiled by the Studsvik library.

   4. Patentlista (with abstracts)
   5. Selektiv litteraturlista (periodical articles)
   } Compiled by the Information Office, Stockholm.

57. *Reaktorn.* (8 issues per year.) AB Atomenergi, Stockholm 9. 1957 and onwards. In Swedish but special issues (e.g. July 1963 issue on the Agesta power station) have long summaries in English.

58. *ASEA Research.* (Irregular.) ASEA, Västeras, Sweden. 1958 and onwards. Editions in English, French and German.
59. *ASEA Journal.* (Monthly.) ASEA, Västeras, Sweden. 1924 and onwards. Published in English, French, German and Spanish editions.

Belgium

60. CENTRE D'ÉTUDE DE L'ÉNERGIE NUCLÉAIRE. *Mol.* (In English.) CEN, 81 rue Belliard, Brussels. 1959.
61. CENTRE D'ÉTUDE DE L'ÉNERGIE NUCLÉAIRE. *Rapport sur l'exercise 1961* (etc.). CEN, 31 rue Belliard, Brussels. 1962 and onwards.
62. *Belgicatom. Bulletin d'Information.* (Bimonthly.) Association Belge pour le Développement Pacifique de l'Énergie Atomique, 35 rue Belliard, Brussels. 1956 and onwards.
63. SOCIÉTÉ BELGE POUR L'INDUSTRIE NUCLÉAIRE. *Reports of the directors.* (In English.) Belgonucléaire, 35 rue des Colonies, Brussels. 1955 and onwards.
64. *ACEC Review.* (Quarterly.) Ateliers de Construction Électriques de Charleroi, Post Box 254, Charleroi, Belgium. 1907 and onwards.
Published in English, French and German editions.
65. *Repertory of the nuclear institutions and industry in Belgium.* (In English.) Groupement Professionel de l'Industrie Nucléaire, 4 rue de la Chancellerie, Brussels. July 1962.
66. CENTRE D'ÉTUDE DE L'ÉNERGIE NUCLÉAIRE. *The development of nuclear energy in Belgium.* (In English.) CEN, 31 rue Belliard, Brussels. 1959.

Netherlands

67. *Atomenergie in der Niederlanden.* (In German.) *In:* Atomwirtschaft, vol. 5, no. 11, November 1963.
A special issue covering the main aspects of atomic energy in Holland in a series of articles of which there are long summaries in English.
68. *The Petten Research Centre of RCN.* (In English.) *In:* Atoomenergie, vol. 4, no. 11, November 1962. (Special issue.)
69. REACTOR CENTRUM NEDERLAND. *The research centre of Reactor Centrum Nederland.* (In English.) External Relations Dept., RCN, 112 Scheveningseweg, The Hague. 1963.
70. REACTOR CENTRUM NEDERLAND. *Summary of activities, January 1962 to June 1963.* (In English.) RCN, 112 Scheveningseweg, The Hague. 1963.
71. *Atoomenergie en Haar Toepassinger (Atomic Energy and its Uses).* (Monthly.) Reflex Publishing Co., 310 Mathernesserlaan, Rotterdam. 1959 and onwards.
Published from 1957 to 1959 under the title "RCN Bulletin". Articles in Dutch with English abstracts.
72. STICHTING VOOR FUNDAMENTEEL ONDERZOEK DER MATERIE. *Jaarvesslagen 1959.* FOM, 4 Lucas Bolwerk, Utrecht. 1960.
73. NETHERLANDS NUCLEAR ENERGY COMMITTEE. *A survey of the nuclear energy organisation in the Netherlands.* (In English.) Netherlands Government Information Service, The Hague. March 1960.
74. *TNO Nieuws.* (Monthly.) TNO Central Office, Koningskade 12, Postbox 297, The Hague. 1946 and onwards. (In Dutch.)
75. *Nuclear industry in the Netherlands.* (In English.) *In:* Atoomenergie, vol. 3, no. 6. June 1961. (Special issue.)

Switzerland

76. *Eidgenossisches Institut für Reaktorforschung, Würenlingen (Swiss Federal Institute for Reactor Research, Würgenlingen). In:* Neue Technik, vol. 2, no. 8, August 1960.
77. Swiss Documentation Service. Cards are issued in three series: (a) Atomic energy (nuclear physics, reactor technology, nuclear engineering, apparatus and instruments. (b) Radiobiology. (c) Nuclear radiation. Abstracts are of periodical articles from the world nuclear literature and are obtainable on subscription from:
Centre de Documentation
Bibliothèque de l'École Polytechnique Fédérale, Leonhardstrasse 33, Zurich 6.
78. *Neue Technik (New Techniques).* (Monthly.) Verlag Neue Technik AG, Badenerstrasse 21, Zurich 4. 1959 and onwards.
Articles and abstracts in English, French and German.
79. *Bulletin de l'Association Suisse pour l'Énergie Atomique.* (Semi-monthly.) A.S.P.E.A., Schauplatzgasse 11, Berne 2. 1959 and onwards.
80. Association Suisse pour l'Énergie Atomique. *Énergie atomique et protection contre les radiations in Suisse: manuel.* (Published in French and German editions.) A.S.P.E.A., Case postale 138, Berne 2.
Issued in collaboration with the Federal Delegate for Atomic Energy.

Norway

81. Netherlands-Norwegian Joint Establishment for Nuclear Energy Research. *Annual report, 1951/52 to 1958/59.* (In English.) JENER, Kjeller, near Lillestrøm, Norway.
82. Institutt for Atomenergi. *The Institutt for Atomenergi, Kjeller and Halden.* (In English.) NFA- Kjeller, near Lillestrøm, Norway. 1961.
83. Institutt for Atomenergi. *Survey of activities, 1948—1960.* (In English.) IFA, P.O. Box 175, Lillestrøm, Norway. 1962.
84. *Teknisk Ukeblad.* (Weekly.) Ingeniörenes Hus, Kronprinsensgt 17, Oslo 9. 1854 and onwards.
Articles in Norwegian with summaries in English.

Denmark

85. Danish Atomic Energy Commission. *Report on the activities of the Danish Atomic Energy Commission.* (In English.) Jul. Gjellerup, 87 Sølvgade, Copenhagen. 1957 and onwards.
Available on exchange from the Library, Danish Atomic Energy Commission, Risø, Roskilde, Denmark.
86. Danish Akademy of Technical Sciences. *DANATOM annual report.* (In English.) DANATOM, 2 Aurehøjvej, Hellerup, Copenhagen. 1956 and onwards.

PART 3

INTERNATIONAL SOURCES OF INFORMATION

Five agencies are described of which only one, the International Atomic Energy Agency, derives its membership from all parts of the world. The rest are European organizations, although in the case of the European Nuclear Energy Agency, Canada and the USA are associate members. One other international organization, the Joint Institute for Nuclear Research at Dubna, near Moscow, which draws its membership from communist countries only, has been described in Chapter 4.

CHAPTER 6

INTERNATIONAL ORGANIZATIONS

INTERNATIONAL ATOMIC ENERGY AGENCY (IAEA)

The IAEA was established in July 1957 under the aegis of the United Nations, with headquarters in Vienna. Its major functions are to encourage research and development in the peaceful uses of atomic energy; to provide member states with needed materials, services and equipment; to foster the exchange of scientific and technical information; to promote the exchange and training of scientists; to establish standards of safety, particularly in the field of radiological protection; and to establish and administer safeguards which will ensure that nuclear materials supplied to member states by the Agency are not used for military purposes.

Organization

A General Conference, consisting of one representative of each of the eighty-three member states, meets annually to elect members of the Board of Governors and to approve the budget and the annual report to the U.N. General Assembly. The Board of Governors is advised by a Scientific Advisory Committee consisting of seven scientists appointed by the Board on the recommendation of the Director-General. The members of this Committee include some of the world's foremost nuclear scientists who serve in an individual capacity, not as national representatives.

The Director-General is appointed by the Board of Governors for a term of four years and has four deputies for (a) administration, (b) training and technical information, (c) research and isotopes, and (d) technical operations, respectively. The permanent staff of the Agency now totals over five hundred persons of many nationalities.

Although the main function of the Agency is to promote international cooperation in atomic energy and particularly to facilitate the flow of information from the highly developed nuclear countries to those not so highly developed, some research is carried out, particularly in the fields of radiobiology, radiological protection and radioactive waste disposal. The Agency has some laboratory facilities at its headquarters and a new research and analytical laboratory has been built at Siebersdorf near Vienna, but the bulk of the research is done through the medium of contracts awarded to existing institutions in various countries.

Publications and Information Services

The work of the Agency is described each year in the *Annual Report of the Board of Governors to the General Conference,* [1] which, with a short supplement, forms the Agency's annual report to the General Assembly of the United Nations. The scientific and technical aspects of the Agency's work are also described in the report [2] which is made each year to the Economic and Social Council of the United Nations. The results of investigations carried out under the research contracts awarded by the Agency are summarized in reports [3] published each year in the Technical Report series.

The publishing programme of the Agency is the responsibility of the Division of Scientific and Technical Information and includes conference proceedings, periodical publications, technical reports, bibliographies and brochures. Each year's publications are listed briefly in the Annual Report of the Board of Governors and all publications (on sale) are described in the Agency's current *Publications Catalogue.* [4]

Each year more than a dozen international conferences and meetings are sponsored by the Agency and the proceedings published. The official languages are English, French, Russian and Spanish and papers are accepted for publication only if they are in one of these languages. In general papers given at conferences are published in the original language (provided it is one of the four named) with summaries in all four. Reports of discussions on papers are printed in the proceedings in English only.

The Agency also issues a number of technical reports which usually arise from meetings of experts convened by the IAEA to study particular aspects of atomic energy. These reports, as well as the proceedings of conferences, may be purchased from the Agency's Distribution and Sales Unit in Vienna, from the United Nations Bookshop in New York or through any of the appointed sales agents in various countries. In general, an agent, such as Her Majesty's Stationery Office in the U.K., which handles United Nations publications will also handle those of the IAEA. A complete list of sales agents is given in each issue of the *IAEA Bulletin,* [5] a quarterly publication designed to disseminate information on the activities of the Agency. The *Bulletin* also gives information about recent publications and summarizes the proceedings of recent IAEA conferences.

The Agency has issued a number of special publications, including technical directories of nuclear reactors and radioisotopes, and a series of booklets on radiological protection and health physics. These are described in the appropriate sections of Part 4. It also published a series of reviews, by prominent scientists, on specific aspects of atomic energy. This series has now been discontinued and replaced by a periodical *Atomic Energy Review* [6] which appears about four times a year.

As a result of exchange arrangements with member states by which the Agency regularly receives unclassified reports and other publications, an extensive library has been established at the Vienna headquarters, providing reference and bibliographical services to the scientific staff of the Agency and to those countries in which atomic energy information services are not well developed. The Documentation Section of the Division of Scientific and Technical Information has issued a number of comprehensive bibliographies on specific aspects of atomic energy, which are listed in the *Publications Catalogue*. [4] The Documentation Section also prepares and issues the *List of References on Nuclear Energy* [7] which appears twice a month in English, and the *List of Bibliographies on Nuclear Energy* [8] which includes bibliographies in preparation as well as those already published.

The *List of References* is based upon the material received in the Agency's Library but does not cover non-nuclear publications of which the Agency receives a great many. These are listed in the *Library Accessions List* [9] which was issued quarterly, but from the beginning of 1964 has appeared monthly, and includes periodic supplements listing the Library's Russian language acquisitions.

One of the Agency's most useful publications is the bi-monthly list of *Conferences, Meetings, Training Courses in Atomic Energy* which is described in Chapter 7; another is the *List of Periodicals in the Field of Nuclear Energy* [10] of which a second edition appeared in 1963. The list contains information on 523 periodicals from 36 countries although only a fraction of these are devoted exclusively to nuclear science and technology.

The Library also administers a collection of films on atomic energy, all of which have been put at the Agency's disposal by member states, and these films are available for loan to organizations who wish to show them for educational, non-commercial and non-profit making purposes. A catalogue [11] of films in the Library has been issued.

EUROPEAN ATOMIC ENERGY COMMUNITY (EURATOM)

Euratom is one of the tangible results of the movement towards Western European unity that began to take practical shape in 1950 when Robert Schumann proposed that the coal and steel industries of France, Germany, Italy, Belgium, Holland and Luxemburg should be placed under a common authority. The European Coal and Steel Community was established in 1951; six years later the same countries signed the Rome treaties by which both the Common Market and Euratom were created.

Western Europe is a region in which conventional fuel resources are no longer adequate to meet requirements and in which imports of fuel have been rising rapidly for some years, a situation which is undesirable both economically and strategically. Forecasts indicate that, if imports are to be kept within reasonable limits, by 1980 over one quarter of the electricity needs of

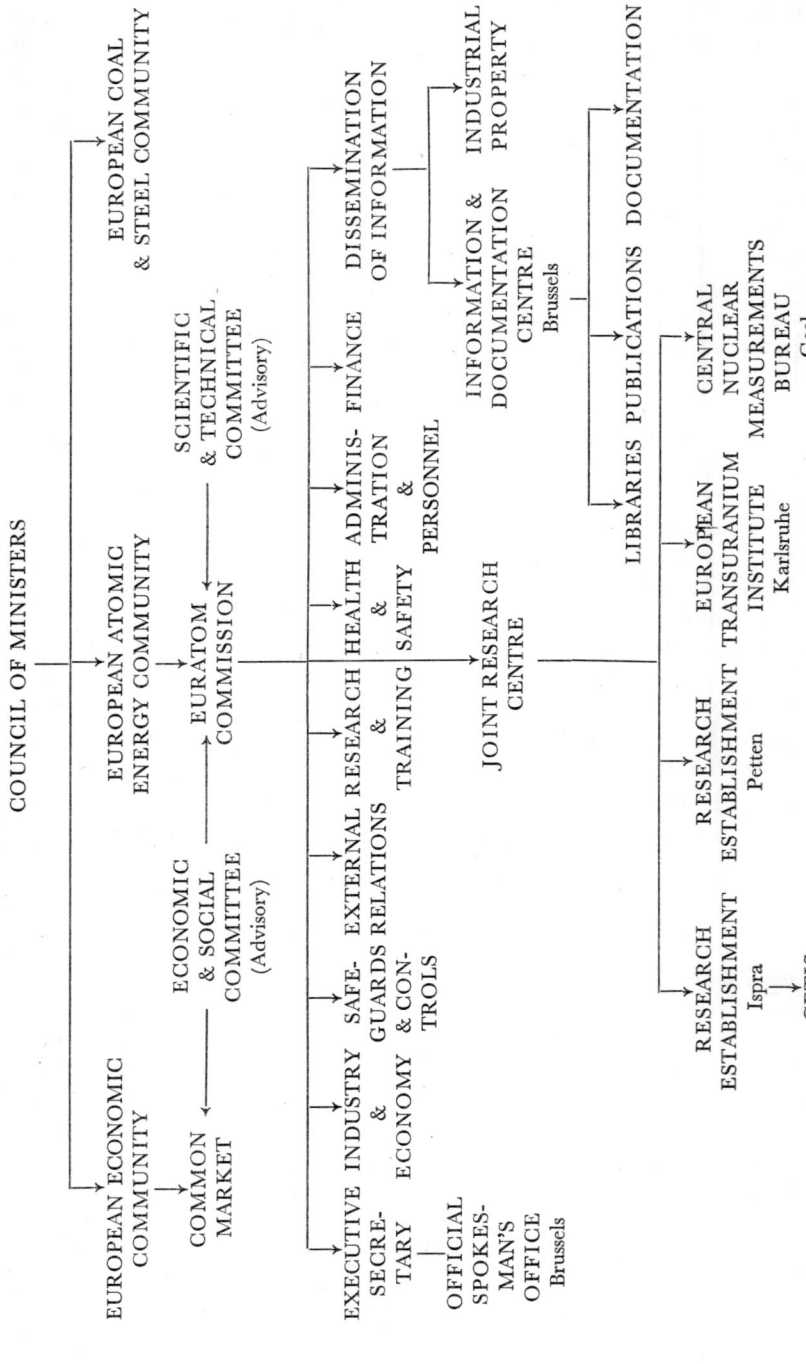

the region will have to be supplied by nuclear power stations. No single country in Western Europe has the financial or technical resources to build a nuclear industry of this size in the time available and only a combined effort will enable the whole of Western Europe to reach the levels likely to be attained by the more advanced nuclear nations.

With something of this in mind, in 1956 the Ministers of Foreign Affairs of the six countries set up a committee of three, Louis Armand, Franz Etzel and Francesco Giordani, who later became known as the "three wise men of Europe", to report on "the amount of atomic energy which can be produced in the near future in the six countries, and the means to be employed for this purpose". Their report, published under the title *Target for Euratom*, [12] is the fundamental document upon which the aims and policies of Euratom are based.

Organization of Euratom

The executive body of Euratom is the Commission, consisting of five members, of different nationalities, who are independent of national governments and sectional interests. The Commission formulates policy, within the terms of the Treaty, administers the research and development programme of the community and reports to the European Parliament. Co-ordination between Euratom's policies and those of the national states is achieved through the Council of Ministers, whose members are representatives of national governments. Since Euratom's funds are provided by national contributions, the Council of Ministers is the final decision-making body as regards new policies, modifications of the Treaty, [13] matters involving financial contributions and the establishment of joint enterprises, although disputes can be referred to the Court of Justice which interprets the treaties and whose judgements are binding on national governments, the Community Executives and others involved.

Research Programme

Since Euratom is looking to the future rather than the immediate present, the essential aim is to gain constructional and operating experience of nuclear power stations rather than to build them immediately in large numbers, and to obtain experience of a number of basic types of power reactor as alternatives to the "proven" types now being built in Britain and America. The main emphasis of the research programme, therefore, is on prototype power reactors, although attention is being given to other aspects.

In order to make the best use of the facilities available, research is conducted in three ways:

(1) At Euratom's own Joint Research Centre.

(2) By awarding contracts for specific tasks to public and private organizations in member countries.

(3) By joint research with non-member countries, either directly as with the U.S.A., or through international organizations such as the European Nuclear Energy Agency.

The Joint Research Centre consists of four research establishments at Ispra (Italy), Geel (Belgium), Petten (Netherlands) and Karlsruhe (Germany) respectively. Ispra is the largest and oldest having been taken over from the Italian Government in 1961, the main project being the ORGEL reactor, a heavy water moderated reactor using an organic liquid as coolant. Ispra is also the site of Euratom's data processing centre, CETIS. The Central Nuclear Measurements Bureau at Geel in Belgium is located close to the Belgian Nuclear Research Establishment at Mol, and will carry out research on instrumentation and measurement techniques. At Petten, in the Netherlands, Euratom has taken over a site adjacent to that of the RCN research centre, together with the high flux materials testing reactor built by RCN who will remain responsible for its operation for four years. The European Transuranium Institute at Karlsruhe has been established alongside the Federal Republic's research centre, and will specialize in the design of plutonium-based fuel elements and the study of other transuranium elements.

Under the terms of an Agreement for Co-operation signed in 1958 a U.S.–Euratom Joint Research Programme has been put into operation, which provides for U.S. assistance to Euratom in the construction of large power reactors and co-operation in related research. The agreement originally envisaged the construction in the Community of a number of power stations of "proven" U.S. design by 1963, but an unexpected easing of the fuel situation in Western Europe has tended to delay the programme and only two of these reactors are under construction at the moment.

A brief account of the organization, aims and facilities of Euratom is contained in a booklet issued in 1964 [14] and a more detailed description will be found in the two special issues [15] of *Neue Technik* for June 1962 and October 1963. The research and development programme is described at some length in the annual reports [16] of Euratom, which cover every aspect of the Community's activities and are available in English.

Information Services

The scientific and technical information services of Euratom are the responsibility of the Information and Documentation Centre in Brussels. The Centre controls the internal dissemination of information within the Commission and its research establishments, as well as external dissemination to the Community and the rest of the world. Its task also includes the fostering of co-operation between Euratom and the national atomic energy documentation centres and between the centres themselves.

The work of the Documentation Centre is based on the Euratom Library in Brussels which is a depository for publications originating from national atomic energy projects inside and outside the Community. Extensive libraries are also being built up at the four research establishments of the Joint Research Centre and these, and their information services, are co-ordinated from Brussels.

Developments in the work of the Centre are reported each year in the annual reports but a more comprehensive account of its aims and facilities is given in a recent Euratom report [17] on regional co-operation within the Community.

The Euratom Commission has encouraged research on data handling and the retrieval and dissemination of information, including mechanical translation, and a good deal of work in this field has been done at the Centre Européen de Traitement de l'Information Scientifique (CETIS) at Ispra. Using an IBM 1401, an information store is now being set up at Brussels, and experiments are being carried out with the object of developing a retrieval system which can be made available, not only to the Commission and its laboratories, but to all potential users in the countries of the Community.

Publications

Euratom has issued a considerable number of scientific and technical reports, the majority of which have been made widely available. Many of these are reprints of journal articles or conference papers but they are all issued in one numerical series and a complete list of those published up to October 1962 has been issued. [18] A similar list is contained in the fifth annual report for 1961–62 and this is supplemented in each succeeding annual report. All reports, except those which are reprints, may be purchased [18] and all are abstracted in *N.S.A.*

Newly published reports are abstracted in *Euratom Information*, [19] a monthly journal which also contains information about patents taken out by the Commission and new research contracts which Euratom has placed with other European organizations. These are exclusive of contracts placed under the U.S.–Euratom Joint Research Programme, the latter being recorded in some detail in the *Quarterly Digest*. [20] The latter journal ceased publication in the early part of 1964 and this information is now included in *Euratom Information*.

The principal publication for disseminating information about the activities of the Community is the *Euratom Bulletin* [21] which appears quarterly. Each issue includes technical articles on some aspects of research and development in the Community together with news of recent events and activities. Euratom also compiles the *Transatom Bulletin* which lists translations of nuclear literature in East European languages and is described more fully in Part 4.

Although the official languages of Euratom are French, German, Italian and Dutch, on the information side English has been adopted as the working language so that nearly all Euratom publications are available in English versions, and English words are used for the thesaurus [22] which is now being compiled in connection with the information retrieval experiment. The main exception to the use of English is the report literature where each document appears in the language of origin and is denoted by the letters f, d, i, n and e after the report number.

The aims and activities of the three Western European Communities overlap considerably and a good deal of information about Euratom is contained in the general publications covering all three organizations, particularly those issued by the European Community Information Service which has offices in all six countries of the Community and in London, Washington and Geneva. Typical is *European Community,* [23] published monthly by the European Community Information Service in London and Washington which contains news items and longer articles on all aspects of the Communities' activities.

EUROPEAN NUCLEAR ENERGY AGENCY (ENEA)

The European Nuclear Energy Agency (ENEA) was formed in February 1958 as a specialized agency of the Organization for European Economic Co-operation (now the Organization for Economic Co-operation and Development), the member countries being Austria, Belgium, Denmark, France, Federal Republic of Germany, Greece, Iceland, Irish Republic, Italy, Luxemburg, Netherlands, Norway, Portugal, Spain, Sweden, Switzerland, Turkey and the United Kingdom. Canada and the U.S.A. are associate members.

The main purpose of the organization is to ensure effective co-ordination of scientific activity in the member countries by the setting up of joint undertakings and by stimulating the exchange of information between the countries so that duplication in the national nuclear research programmes is avoided wherever possible. Another important function of ENEA is to create the economic and social conditions which are necessary for a rapidly developing nuclear industry in Europe.

ORGANIZATION

A Steering Committee for Nuclear Energy was set up in 1956 to submit proposals regarding joint action in the nuclear field to the Council of OEEC, as it then was. When the Agency was established, the Steering Committee became the governing body, responsible to the Council of OEEC. A small permanent secretariat has been formed at OECD headquarters in Paris and a number of working parties and study groups have been brought into being to examine particular aspects of atomic energy development.

INTERNATIONAL ORGANIZATIONS

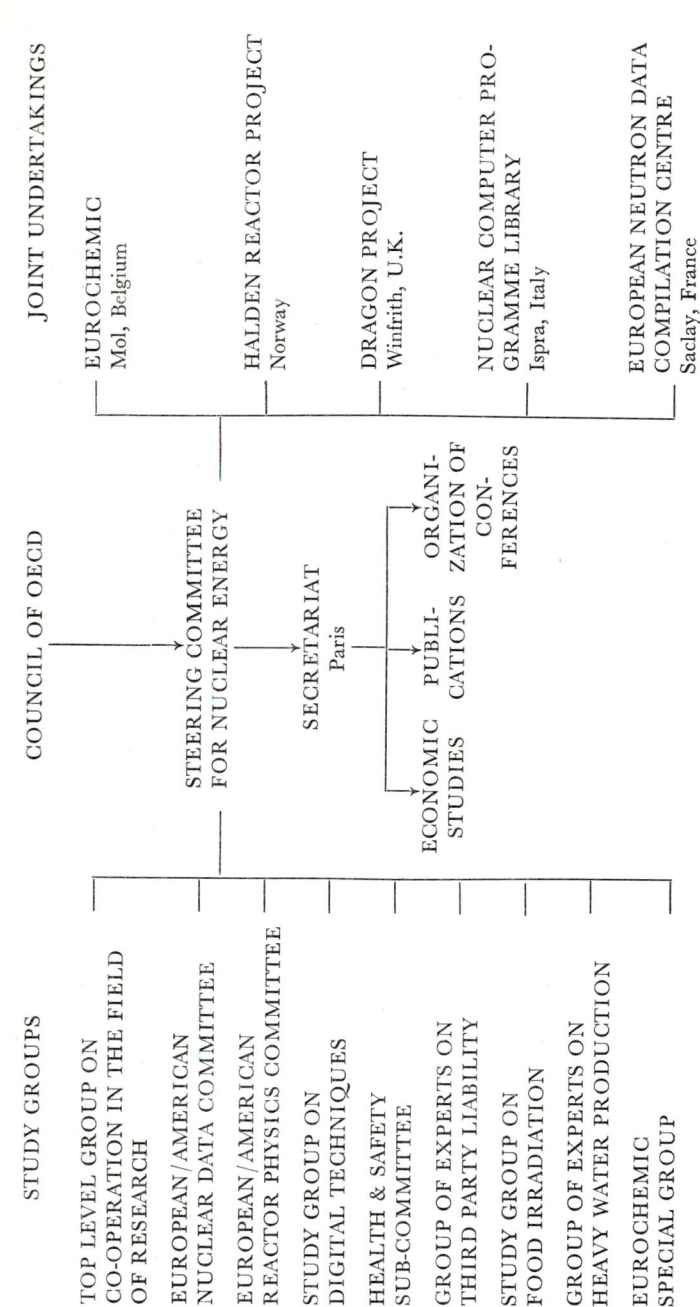

ORGANIZATION OF ENEA

The organization and functions of the Agency are set out in the *Statute of the Agency* [24] which was published in Britain as a white paper in 1958. The text of this document together with the agreements setting up the joint undertakings was also published by OEEC in 1961. [25] A short booklet [26] describing the structure and function of ENEA and including a short account of the first joint undertakings was issued in 1958.

Activities

The most important activity of the Agency has been the establishment of several joint undertakings in which member countries can participate. Research and development projects set up in this way are the subject of separate agreements and the countries taking part vary from one undertaking to another. At present three such projects exist and two others are being set up. In chronological order they are:

European Company for Chemical Processing of Irradiation Fuels (Eurochemic)

The Convention establishing Eurochemic was signed in 1957 and came into force in July 1959, the signatory countries being Austria, Belgium, Denmark, France, Germany, Italy, Netherlands, Norway, Portugal, Spain, Sweden, Switzerland and Turkey. The objectives are to build and operate a plant for processing irradiated fuels from European reactors and to carry out research towards the development of economic methods of fuel processing. The plant and research laboratories are being built on a site adjacent to the Belgian research centre at Mol.

A detailed description of the scientific and technical activities of Eurochemic is contained in the *First Activity Report* [27] published in 1963, which also contains a list of the technical reports issued by Eurochemic up to March 1962.

Halden Reactor Project

The Halden Project was formed as a result of an agreement signed in 1958 by Austria, Denmark, Euratom, Norway, Sweden, Switzerland and the United Kingdom. The main purpose of the project is to enable participating countries to obtain practical experience in the design and operation of boiling water reactors in which heavy water is used as both coolant and moderator. The reactor, situated at Halden in Norway, is owned by the Norwegian Institutt for Atomenergi, who began construction of the reactor in 1957 before the agreement was signed, and are still responsible for its operation. The joint programme of research and experiment on the reactor is sponsored by the participating countries and operates through the Halden Committee on which all the participants are represented. The initial research programme came to an end in December 1962 but was extended until June 1964. In December 1963 a new agreement was signed as a result of which the research programme was extended to December 1966. The participating countries in the new agreement are Denmark, Finland, Netherlands, Norway, Sweden, Switzerland and the United Kingdom.

The experimental programme is described at some length in the *Annual Reports* [28] which also contain lists of the reports issued during the year under review. A brief description [29] of the organization, aims and main facilities of the Halden Project was issued by ENEA in 1962.

High Temperature Reactor Project (Dragon)

The agreement for the design, construction and operation of a high temperature, gas-cooled reactor was signed in March 1959, the participating countries being Austria, Denmark, Euratom, Norway, Switzerland, Sweden and the United Kingdom. The agreement provided for a programme of research and experiment with this type of reactor, extending for a period of five years, but in view of the time required to design and build the reactor, which did not come into operation until 1964, the agreement has been extended for a further period of three years.

The reactor is located on the site of the UKAEA Atomic Energy Establishment at Winfrith in Dorset, and is described in the *Annual Reports* [30] on the project which also give an account of the experimental programme. A number of technical reports have been issued including Dragon Project Reports and Contractors' Reports but few of these are publicly available and the publications listed in the annual reports refer mainly to papers published in scientific journals.

Nuclear Computer Programme Library

The project to set up a Computer Programme Library at the Euratom Centre at Ispra in Italy was approved by the Steering Committee of ENEA in November 1963. The aim is to improve collaboration between originators and users of computer programmes within the framework of ENEA.

European Neutron Data Compilation Centre

A proposal to establish a Neutron Data Compilation Centre at Saclay in France has been approved by the Steering Committee. The Centre will compile and maintain indexes of bibliographical references to publications containing information on neutron data and circulate this information in a usable form. Experimental neutron data produced in Europe and elsewhere will be collected, classified and made available on request.

Apart from these specific projects the Agency has a number of committees and study groups working on other aspects of atomic energy including nuclear ship propulsion, heavy water production, reactor physics and various economic studies. All these are described in the Agency's annual report [31] which also summarizes the progress of the joint undertakings described above.

In order to further the purpose of creating conditions necessary for the development of an expanding nuclear industry in Europe, both ENEA and its parent body OECD have sponsored a number of conferences and symposia of which the most important have been the three conferences on *Nuclear Energy for Management,* the proceedings of which have been published under the title *Industrial Challenge of Nuclear Energy.* [32] The conferences were concerned mainly with the economic, financial and marketing aspects of atomic energy but the proceedings contain some useful information on the nuclear programmes of the European countries.

The Agency has not overlooked the importance of education and training in this field and has sponsored a number of special courses in collaboration with various national atomic energy organizations. In addition, it publishes an annual *Catalogue of Courses on Nuclear Energy and Technology* [33] which covers a wide range of training courses in sixteen countries.

EUROPEAN ORGANIZATION FOR NUCLEAR RESEARCH (CERN)

It was observed in Chapter 1 that one of the essential requirements for research in fundamental nuclear physics is the availability of powerful particle accelerators of which the largest are so complex and expensive as to be beyond the financial or technical resources of many countries. Kowarski, in his account [34] of the origins of CERN says that, "The early history of CERN is that of an encounter between two drives which became operative in Europe immediately after the war; the scientists' search for new ways of acquiring large-scale equipment and the statesman's search for domains of common interest in which a joint effort could be made to produce tangible manifestations of European unity." Be this as it may, the price of a place in the front rank of high energy physics research is high, and this is clearly a field in which international co-operation offers considerable advantages.

Organization and Facilities

A Council of Representatives of European States was formed in 1952 for the purpose of planning an international laboratory and organizing other forms of co-operation in nuclear research. Under the sponsorship of UNESCO the Council (abbreviated in French to "Conseil Européen pour la Recherche Nucléaire" from which the initials CERN are derived) appointed a Secretary General and set up "study groups" to prepare a scientific programme for the new organization which was to be located at Meyrin, near Geneva.

In July 1953, a Convention establishing a European Organization for Nuclear Research, to carry out fundamental research in nuclear physics and associated fields, was signed in Paris, the signatory countries being France, Belgium, Denmark, Federal Republic of Germany, Greece, Italy, Netherlands, Norway, Sweden, Switzerland, United Kingdom and Yugoslavia, and the first meeting of the permanent Council took place in October 1954.

INTERNATIONAL ORGANIZATIONS

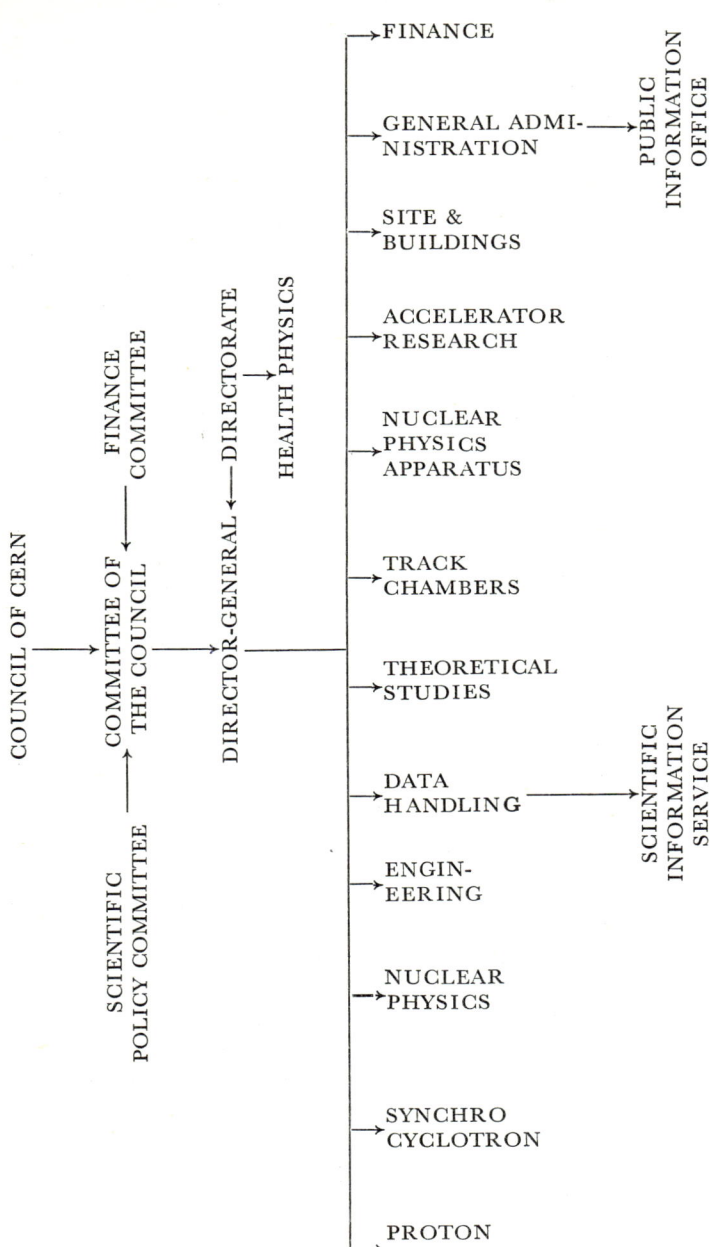

The Council of CERN, which is the governing body, consists of not more than two delegates from each member state and meets at least once a year. The Council operates through three standing committees, the Committee of Council which meets several times a year, the Scientific Policy Committee which advises on scientific questions and the Finance Committee which has wide powers over the financial administration of the organization. Executive authority lies with the Director-General who controls the twelve major divisions of CERN with the aid of a Directorate of four members.

The work of the Laboratory, which is described each year in the *Annual Report* [35] of the organization, centres around the two large particle accelerators, a 600 MeV synchro-cyclotron which came into operation in 1957 and a 28,000 MeV proton synchroton which started operating in November 1959. The early years of CERN were devoted mainly to the design and construction of the accelerators, but since 1959 the emphasis has shifted to nuclear and high energy physics. The machines have been designed in such a way that a large number of experiments can be performed simultaneously, but the demand for machine time is so great that the experimental programme has to be very carefully planned if the machines are to be used effectively. Experiments are usually conducted by mixed teams of CERN staff members and visiting scientists and a number of Experiments Committees have been set up to co-ordinate the various experimental programmes and to facilitate co-operation between the visiting teams and the resident staff.

Information Services

The primary task of the Scientific Information Service of CERN is to provide a library and information service for the staff of the Laboratory although, in the field of high energy physics, it is willing to help other potential users as far as its resources will allow. The Service is also concerned with the production and distribution of scientific and technical reports and the proceedings of conferences held at CERN. The work of the Service is described in some detail in the *Guide to the CERN Library and Related Services* [36] which, although designed primary for CERN staff, contains a good deal of useful information for anyone interested in this field.

The CERN Convention states explicitly that the results of research shall be made generally available and this is done by publication in scientific and technical journals and conference proceedings or, if the information is of a preliminary or tentative nature, by the issuing of technical reports. The latter are not available through commercial channels but are sent on exchange to other organizations and are made available to individual scientists on request, as long as stocks last. A list [37] of reports and papers issued up to 1959 has been compiled and this is supplemented by six-monthly lists of reports and reprints which are cumulated at regular intervals. These lists can be obtained from the Scientific Information Service.

CERN reports are sent to the UKAEA, the USAEC and many other official atomic energy projects, in return for which the CERN Library receives most of the publications issued by these projects including a good deal of material from Eastern Europe. A number of bibliographies have been published and a *Catalogue of Periodicals in the CERN Library* [38] which forms a useful guide to the periodical literature of nuclear science.

Important events in which CERN is concerned are the subject of press releases which are issued by the Public Information Office. This office maintains a collection of photographic prints and films depicting the work of the Laboratory, and publishes the *CERN Courier,* [39] a house journal which appears each month in both English and French editions and contains technical articles as well as news of events at CERN.

EUROPEAN ATOMIC FORUM (FORATOM)

FORATOM is a non-profit-making association of atomic industrial associations, set up in 1960 to promote the activities of its members at international level, through the organization of meetings and conferences, the creation of services for keeping members in touch with nuclear developments and the formation of study groups to examine particular problems.

Its active members are the national atomic forums of eleven European countries, Austria, Belgium, France, Federal Republic of Germany, Italy, Luxemburg, Netherlands, Norway, Portugal, Spain and Switzerland and its headquarters is housed in the offices of the French organization, Association Technique pour l'Énergie Nucléaire (ATEN) in Paris.

The principal feature of FORATOM is the international congress which, in principle, is held every two years, the first one taking place in Paris in September 1962 on "Conditions for the creation and development of a European nuclear industry." This congress was reported at some length in *ATEN Bulletin* [40] which up to 1962 devoted considerable space to FORATOM developments. Since then FORATOM has published its own journal [41] which gives full details of its work and programmes.

REFERENCES

1. INTERNATIONAL ATOMIC ENERGY AGENCY. *Annual report of the Board of Governors to the General Conference.* 1958 and onwards. I.A.E.A., Kärntnerring 11, Vienna 1, Austria. (In English.)
2. INTERNATIONAL ATOMIC ENERGY AGENCY. *Annual report to the Economic and Social Council of the United Nations.* 1960 and onwards. I.A.E.A., Vienna 1. (In English.)
3. *I.A.E.A. research contracts.* First Annual Report 1961. *Technical Reports Series No. 4.* Second Annual Report 1962. *Technical Reports Series No. 9.* Third Annual Report 1963. *Technical Reports Series No. 16.* I.A.E.A., Vienna 1. Published in English, French, Russian and Spanish versions.

4. INTERNATIONAL ATOMIC ENERGY AGENCY. *Publications in the nuclear sciences, 1964.* I.A.E.A., Vienna 1. March 1964.
5. *International Atomic Energy Agency Bulletin.* (Quarterly.) I.A.E.A., Vienna 1. April 1959 and onwards.
   Published in English, French, Russian and Spanish editions.
6. *Atomic Energy Review.* (Irregular.) I.A.E.A., Vienna 1. 1963 and onwards.
   Papers in either English, French, Russian or Spanish with summaries in all four languages.
7. *List of references on nuclear energy.* (Semi-monthly.) I.A.E.A., Vienna 1. 1959 and onwards. (In English.)
   List of books, reports and other documents arranged under broad subject headings. An index to authors and report numbers covering the years 1959—1961 was issued in 1963.
8. *List of bibliographies on nuclear energy.* (Irregular.) Vol. 1, no. 1, June 1960. Vol. 1, no. 2, March 1961. Vol. 2, no. 1, October 1962. I.A.E.A., Vienna 1. (In English.)
   Arranged under broad subjects. Each volume has an author index.
9. *I.A.E.A. Library accessions list.* (Quarterly.) Library, I.A.E.A., Vienna 1. 1958 and onwards. (In English.)
10. *List of periodicals in the field of nuclear energy.* Revised edition. (In English.) I.A.E.A., Vienna 1. 1963.
    A list compiled on the basis of information received from member states. For each journal the following information is given: Title; abbreviation; publisher; sponsor; editor; periodicity; vols. published; type of information; language; circulation; subscription rate; address for orders. Includes a supplementary list of journals on space research and technology.
11. *Films on the peaceful uses of atomic energy available through the International Atomic Energy Agency. Vienna.* Catalogue no. 2. I.A.E.A., Vienna 1. 1963. (In English.)
12. ARMAND, L., ETZEL, F. AND GIORDANI, F. *A target for Euratom: report submitted by Mr. Louis Armand, Mr. Franz Etzel and Mr. Francesco Giordani at the request of the governments of Belgium, France, German Federal Republic, Italy, Luxemburg and the Netherlands.* May 1957. (In English.)
13. *Treaty establishing the European Atomic Energy Community (Euratom), Rome, 25 March, 1957.* London, H.M.S.O., 1962.
    This is an English translation of the Treaty which must not be regarded as an official or authentic text. The latter exists only in the French, German, Italian and Dutch languages.
14. EURATOM COMMISSION. *Euratom: the European Atomic Energy Community.* European Community Press and Information Service, 23 Chesham St., London, S.W.1. 1964. (In English.)
15. (a) *Organisation et activités de la Communauté Européenne de l'Énergie Atomique (Euratom). In:* Neue Technik, vol. 4, no. 6, June 1962.
    (b) *Euratom: Communauté Européenne de l'Énergie Atomique. In:* Neue Technik, vol. 5, no. 10, October 1963. Individual articles in French, German or English.
    (a) contains a detailed account (in German) of the Euratom Information and Documentation Centre.
16. EUROPEAN ATOMIC ENERGY COMMUNITY. *General report on the activities of the Community.* 1958 and onwards. Euratom Commission, 51—53 rue Belliard, Brussels. (English version available.)
17. BREE, R. *Regional co-operation in the field of scientific and technical information within the European Community.* (EUR 494 e.) (In English.) Euratom, Information and Documentation Centre, 1964. Available from: Presses Academiques Europénnes, 98 Chaussée de Charleroi, Brussels 6.

18. EUROPEAN ATOMIC ENERGY COMMUNITY. *List of the scientific and technical reports published by the Commission.* Euratom, Information and Documentation Centre, 51—53 rue Belliard, Brussels.
Includes reports issued up to 15 October 1962 or in the press on that date. Reports, other than reprints, may be purchased from: Presses Academiques Européennes, 98 Chaussée de Charleroi, Brussels 6, *or* Office Central de Vente des Publications des Communautés Européennes, 2 place de Motz, Luxemburg.
19. *Euratom Information.* (Monthly.) Verlag Handelsblatt G.m.b.H., Postfach 1102, 4 Düsseldorf, Federal Republic of Germany. 1963 and onwards.
In English with contents list also in French, German, Italian and Dutch. Abstracts are in the language of the article and in English if this is not the language of origin. Each issue has personal and corporate author indexes, and an annual index is published.
20. *The Joint Research and Development Program Quarterly Digest.* (Quarterly.) Euratom, Information and Documentation Centre, 51—53 rue Belliard, Brussels. 1960 and onwards. (In English.)
21. *Euratom Bulletin of the European Atomic Energy Community.* (Quarterly.) A. W. Sijthoff, Postbox 26, Leiden, Netherlands. 1962 and onwards.
Published in English, French, German, Italian and Dutch editions.
22. EUROPEAN ATOMIC ENERGY COMMUNITY. *Euratom Keyword thesaurus.* (EUR 500 e.) (In English.) Euratom, Information and Documentation Centre, 51—53 rue Belliard, Brussels.
23. *European Community.* (Monthly.) European Community Information Service, 23 Chesham St., London, S.W.1., *or* 235 Southern Building, Washington 5, D.C. (In English.)
24. FOREIGN OFFICE. *Decision of the Council of the Organisation for European Economic Co-operation establishing a European Nuclear Energy Agency and Convention on the Establishment of a Security Control in the field of Nuclear Energy, Paris, December 20, 1957.* London, H.M.S.O., February 1958. Cmnd 357.
25. ORGANIZATION FOR EUROPEAN ECONOMIC CO-OPERATION. EUROPEAN NUCLEAR ENERGY AGENCY. *Statute of the Agency. Convention on security control. Acts constituting the joint undertakings (Eurochemic, Halden, Dragon). 1957—1960.* OECD, 2 rue André-Pascal, Paris 16ᵉ. 1961.
26. EUROPEAN NUCLEAR ENERGY AGENCY. *The OEEC European Nuclear Energy Agency. Structure and functions: first joint undertakings.* OEEC, Paris. 1958.
27. EUROPEAN COMPANY FOR THE CHEMICAL PROCESSING OF IRRADIATED FUELS (EUROCHEMIC). *First activity report. 1959—1961.* ENEA, 2 rue André-Pascal, Paris 16ᵉ. 1963.
28. OECD HALDEN REACTOR PROJECT. *Annual report.* 1959 and onwards. ENEA, 2 rue André-Pascal, Paris 16ᵉ.
29. OECD HALDEN REACTOR PROJECT. *Halden heavy boiling water reactor.* ENEA, 2 rue André-Pascal, Paris 16ᵉ. 1962.
30. OECD HIGH TEMPERATURE REACTOR PROJECT (DRAGON). *Annual report.* 1959/60 and onwards. ENEA, 2 rue André-Pascal, Paris 16ᵉ.
31. EUROPEAN NUCLEAR ENERGY AGENCY. *Report on the activities of the Agency.* 1958 and onwards. ENEA, 2 rue André-Pascal, Paris 16ᵉ.
32. ORGANIZATION FOR EUROPEAN ECONOMIC CO-OPERATION. *The industrial challenge of nuclear energy.*
   1. *Proceedings of the first Information Conference on Nuclear Energy for Management, Paris, April 1957.* Published by the European Productivity Agency of OEEC, 1957.
   2. *Proceedings of the second Information Conference on Nuclear Energy for Management, Amsterdam, June, 1957.*

3. *Proceedings of the Stresa Conferences, May 1959.* Published in four parts by ENEA, 1959.
OECD, 2 rue André-Pascal, Paris 16ᵉ.

33. EUROPEAN NUCLEAR ENERGY AGENCY. *Catalogue of courses on nuclear energy and technology in the European countries of OECD.* 1957/58 and onwards. ENEA, 2 rue André-Pascal, Paris 16ᵉ. (Annual.)
Includes courses in universities, higher technical colleges and research centres in sixteen countries.

34. KOWARSKI, L. *An account of the origin and beginnings of CERN.* CERN, Geneva, 1961. (CERN 61—10.)
Contains the text of the Convention establishing CERN.

35. EUROPEAN ORGANIZATION FOR NUCLEAR RESEARCH. *Annual report.* 1955 and onwards. CERN, Geneva 23.
Includes lists of reports and journal articles published each year and a chapter on the work of the Scientific Information Service.

36. EUROPEAN ORGANIZATION FOR NUCLEAR RESEARCH. *Guide to the CERN Library and related services.* Scientific Information Service, CERN, Geneva 23. 1962. (CERN Bibl. 6.)

37. EUROPEAN ORGANIZATION FOR NUCLEAR RESEARCH. *Index of scientific publications, 1955—1959.* Scientific Information Service, CERN, Geneva 23. 1960.
With author and subject indexes.

38. EUROPEAN ORGANIZATION FOR NUCLEAR RESEARCH. *Catalogue of periodicals in the CERN Library.* Scientific Information Service, CERN, Geneva 23. 1960. (CERN Bibl. 5.)
In part 1 the journals are arranged alphabetically by title: in part 2 they are arranged in subject order by UDC.

39. *CERN Courier.* (Monthly.) Public Information Office, CERN, Geneva 23. August 1959 and onwards.
Published in English and French editions and issued without charge to those interested in the construction and use of particle accelerators or in the progress of nuclear physics.

40. Le Congrès de FORATOM. (Preprints of papers delivered at the Congress.) *In: Bulletin d'Information A.T.E.N.,* no. 36, July/August 1962.
Le Congrès de FORATOM. (Reports on the Congress.) *In: Bulletin d'Information A.T.E.N.,* no. 37, September/October 1962.

41. *FORATOM Informations: Bulletin d'Information du Forum Atomique Européan.* FORATOM, 26 rue de Clichy, Paris 9ᵉ. 1962 and onwards.

PART 4

PUBLISHED SOURCES OF INFORMATION

The published literature of atomic energy is scattered through a very large number of books, journals, conference proceedings, pamphlets, catalogues, specifications and other documents since there are few fields of science and technology on which it has not made some impact. Many of these documents are primarily concerned with other subject fields and are excluded from this survey. Mention is made only of publications which are concerned entirely, or mainly, with those aspects of atomic energy described in Chapter 1, and even here a good deal of selection has been necessary.

For the sake of clarity, the field has been divided into five sections although, in practice, there is some overlap. Chapter 7 describes the general publications on the subject and is sub-divided, for convenience, on the basis of the form in which the document is published. Chapter 8 deals with nuclear and high-energy physics, Chapter 9 with nuclear power and nuclear engineering including raw materials and processing, and Chapter 10 with ionizing radiation and radioisotopes. The last chapter is devoted to controlled nuclear fusion and plasma physics.

CHAPTER 7

ATOMIC ENERGY IN GENERAL

BIBLIOGRAPHIES AND ABSTRACTING SERVICES

The most up-to-date and comprehensive bibliographies of atomic energy literature are those issued by the major national projects, particularly the USAEC, the UKAEA and the French CEA, and the most important of these have been described in previous chapters. Earlier material is covered fairly comprehensively in the two-volume bibliography [1] compiled by the United Nations Atomic Energy Commission which, with its supplements, indexes international literature published before 1954. A more selective bibliography, covering the first twelve years of atomic energy and not previously mentioned, is the *Selected Readings on Atomic Energy*, [2] issued by the USAEC in 1958, although most of the items included are of American origin. British publications up to 1958 are described in two articles [3] which appeared in *British Book News* in 1958, although this is by no means a comprehensive list.

The most important continuing bibliography of atomic energy is, of course, *Nuclear Science Abstracts* [4] which has already been mentioned. This journal began publication in 1946 as *Abstracts of Declassified Documents* and changed its title in 1947 when its scope was broadened to include material other than USAEC reports. It now covers report literature issued by government agencies, universities and industrial research organizations in many countries, as well as journal articles, conference papers, patent specifications, books and pamphlets selected from world literature. Abstracts, in English, are arranged under broad subject headings and each issue contains author, subject and report number indexes which are cumulated quarterly, half-yearly, annually and five-yearly. The annual report number index actually lists all report numbers announced since 1946 and gives information about the public availability of each report. In many ways *N.S.A.* is a model of all an abstracting service should be. It is up-to-date (on average abstracts appear about three to four months after publication of the original), extremely well indexed with frequent cumulations, accurate, attractively produced, and made available to the public at a comparatively low cost. It indexes just over 35,000 items each year and a list of the periodicals from which articles have been abstracted appears in each annual subject index.

Whereas in *N.S.A.* the coverage of report literature is perhaps better than that of journal articles, the opposite is true of *Science Abstracts*. [5] This is not, of course, a journal devoted entirely to nuclear science but it does include a great deal of atomic energy literature. *Section A—Physics Abstracts* includes atomic and nuclear physics, plasma physics, and related fields, the abstracts being arranged under headings which follow the order of the Universal Decimal Classification, although the U.D.C. notation is not actually used. *Section B—Electrical Engineering Abstracts* includes nuclear engineering and nuclear power, particle accelerators, thermonuclear devices, radiation detectors and other instrumentation and is arranged by U.D.C. using the notation of the classification. It is the present policy of *Science Abstracts* not to include report literature but journal coverage, particularly of European journals is extremely good. Annual author and subject indexes are issued separately for both sections. *Science Abstracts* is published by the Institution of Electrical Engineers in association with the Institute of Physics and the Physical Society, with financial support from the American Institute of Physics, and over 40,000 items are now being abstracted each year. Attempts are being made to improve the indexing of the journal by the introduction of computer indexing systems similar to those used in the production of *N.S.A.*

Chemical aspects of nuclear science are covered by *Chemical Abstracts* [6] probably the world's most comprehensive abstracting service in the scientific field. The journal also has a section on nuclear physics but, even so, the proportion of abstracts on atomic energy and its aspects in this journal is very small and its high cost will put it beyond the resources of many libraries. Abstracts are arranged under broad subject headings with an author index in each issue, and cumulated author and subject indexes are published annually and at five-yearly intervals.

It has already been mentioned that many U.S. Government agencies, other than the USAEC, issue reports of interest to nuclear scientists and engineers and that many of these are sold by the Office of Technical Services of the U.S. Department of Commerce. These reports are abstracted in *U.S. Government Research Reports,* [7] a semi-monthly journal which announces reports issued by the U.S. Army, Navy and Air Force, the USAEC, and other agencies of the Federal Government. The first section of each issue, entitled *Technical Abstract Bulletin*, is compiled by the Defence Documentation Center and includes reports resulting from research carried out under defence contracts. The second section, entitled *Non-Military and Older Military Research Reports,* includes reports issued by the USAEC and other agencies and older military reports which have been acquired by OTS in response to specific requests from industry. Within each section abstracts are arranged under broad subject headings and serial number and subject indexes (but no author indexes) are provided in each issue. Cumulated corporate author, subject and report number indexes are issued separately. For about sixteen months a computer-produced "keywords" index was distributed without charge to subscribers to *U.S. Government Research Reports* as an experiment to determine the potential value of this kind of indexing, but this service was discontinued in October 1963 with the comment that, "a survey of recipients revealed that the majority did not consider it sufficiently useful to pay a subscription fee for its continuance".

On the engineering side two publications are worthy of note. *Nuclear Engineering Abstracts* [8] covers technical literature and reports dealing mainly with the engineering and technological aspects of atomic energy. In many respects this journal is unique among abstracting journals. The abstracts are long and detailed and in some cases amount almost to an analysis of the information contained in the original documents. If, for example, three papers are published on a subject at the same time, the abstract deals with all three together, picking out the salient points of each and comparing them. Such an approach is ideal for the user who wishes to keep informed of current information, although it is less useful for retrospective searching. Unfortunately this kind of abstracting requires a particular combination of skills which is rare and some difficulty has been found in assembling a suitable editorial team. As a result there have been considerable delays in publication so that the advantages of the format as a vehicle for current information have been obviated.

Engineering Index, [9] published under the auspices of four leading American engineering societies, covers all aspects of engineering including nuclear and chemical engineering. Up to 1961 the *Index* was issued only in annual volumes; since then it has also appeared monthly with author and subject indexes arranged as in the annual volumes. A weekly card service is also available by which subscribers can receive cards relating to items in any of 249 subject categories, the entries being the same as those that will later appear in the monthly and annual issues.

All the abstracting journals so far mentioned are published in English and although the majority of them seek to survey the world nuclear literature, there is a natural tendency to abstract English language publications more comprehensively than those in other languages. For example, with the possible exception of *N.S.A.*, none of them gives adequate coverage of Russian literature and here recourse must be had to the appropriate sections of *Referativnyi Zhurnal* which was described in Chapter 4. Two journals which give greater emphasis to publications in other European languages are *Bulletin Signalétique*, [10] published by the Centre National de la Recherche Scientifique in Paris, and *Zentralblatt für Kernforschung und Kerntechnik*, [11] compiled by the Deutschen Akademie der Wissenschaften in Berlin. The *Bulletin Signalétique* is issued in twenty-two separate sections of which the most useful in the nuclear field is Section 5—Nuclear Physics. Entries are arranged in classified order with an author index in each issue. Annual author and subject indexes are issued and nearly 7000 items are abstracted each year. *Zentralblatt für Kernforschung und Kerntechnik* indexes over 10,000 items a year from the world's nuclear literature with emphasis on nuclear physics and engineering. Entries are arranged in U.D.C. order and annual author and subject indexes are provided. This journal is particularly useful for its coverage of German and Eastern European literature.

Turning now to bibliographical services without abstracts, mention has already been made of the *List of References on Nuclear Energy* published by the International Atomic Energy Agency. A more comprehensive publication is the *Indexed Bibliography* [12] published by the Gmelin Centre for Atomic Energy Documentation. This was first issued in two sections: series A, which covered reports from national projects (mainly those of the USAEC), and series B, which included references to conference papers, dissertations, patents, books and pamphlets. In 1962 these were combined into one publication, series A–B. A subject index is published separately with each issue of the bibliography and these are cumulated annually. Also published by the AED Information Services is series C, *Bibliographic Review of Selected Subjects*, [13] each issue of which is a bibliography of current literature on a specific aspect of atomic energy. These bibliographies are compiled for AED by various specialist organizations in Germany and elsewhere, and each issue has an author and subject index.

A great many abstracting and indexing periodicals include one or more aspects of atomic energy in their coverage and some of these, dealing with specialized topics, will be mentioned in later chapters. The most up-to-date guide to these is the *Guide to the World's Abstracting and Indexing Services in Science and Technology*, [14] prepared by the Science and Technology Division of the Library of Congress for the National Federation of Science Abstracting and Indexing Services. This publication lists 1855 scientific indexing periodicals originating in 40 countries, of which more than half come from only five countries, U.S.A., Britain, Western Germany, France and the U.S.S.R. The subject fields covered by each journal are indicated in some detail so that this publication is particularly helpful in locating appropriate sources.

DIRECTORIES AND HANDBOOKS

Reference has already been made, in Part 2, to a number of national guides and directories to atomic energy organizations in particular countries. In addition to these there are one or two directories covering nuclear organizations in a number of countries and a few general directories which include a good deal of nuclear information.

The most comprehensive guide to national and international atomic energy organizations is the *World Nuclear Directory*, [15] which gives detailed information about atomic energy authorities and commissions, associations, societies, government departments, university institutes and industrial firms in 76 countries together with the names of their senior staff. An appendix lists the major nuclear periodicals published in each country with the names of their editorial staff. In Britain, many organizations which are able to provide information in the atomic energy field are listed in the *Aslib Directory*, [16] although this publication and its supplements contains no information later than 1960. A similar publication, [17] covering the United States has been issued more recently by the National Science Foundation and lists some agencies not covered by the *World Nuclear Directory*. Russian scientific organizations and laboratories are described in the *Directory of Selected Scientific Institutions in the U.S.S.R.*, [18] prepared by Battelle Memorial Institute for the National Science Foundation. This directory contains short descriptions of the work of each institute and lists the senior staff when these are known.

For biographical information the most compact source is *Who's Who in Atoms*, [19] the third edition of which has over 17,000 entries representing individuals from seventy countries. The amount of information given varies considerably from one person to another, and sometimes it is a little out-of-date, but there are comparatively few omissions. Biographical information on nuclear scientists in the English-speaking world will also be found in *American Men of Science* [20] which has nearly 100,000 entries in the volumes relating to physical and biological science, and the *Directory of British Scientists* [21] which is almost entirely confirmed to those possessing university degrees in science and excludes many engineers and others who hold only professional qualifications.

The Atomic Energy Deskbook, [22] sponsored by the USAEC, combines the functions of directory and handbook. Arranged as an encyclopaedia its main object is to provide background information by which the status of atomic energy development in America and other countries can be judged, with some emphasis on the legal and political framework within which nuclear research and development is conducted. Details of atomic energy organizations, national programmes and energy resources are given for many countries, as well as a good deal of technical information, although it is primary concerned with nuclear science and technology in the U.S.A. A useful feature is a selected reading list of about 250 entries grouped under 29 subject headings.

There are a number of handbooks concerned solely with the scientific and technical aspects of atomic energy. The *Nuclear Engineering Handbook* [23] is, in spite of its title, a compilation of basic data on nuclear physics, accelerators, reactor physics, reactor engineering, and the separation and production of isotopes, and has a long list of references at the end of each chapter. A similar compilation is the *Nuclear Handbook*, [24] intended mainly for nuclear physicists and covering physics, chemistry, radiation detection and instrumentation. Jefferson's *Handbook of the Atomic Energy Industry* [25] is a broad introduction to nuclear energy covering organization and development, reactor engineering and reactor materials, isotopes, radiological protection and waste disposal and is written at a level suitable for the non-specialist and intelligent layman. A much more elaborate publication which covers every aspect of atomic energy is *Génie Atomique*, [26] published in four volumes and based on courses given at the Institut National des Sciences et Techniques Nucléaires at Saclay. It is issued in looseleaf form so that it can be kept up-to-date more easily and is very well documented.

Nuclear data can also be found in a number of other handbooks. The *Hand-*

book of Chemistry and Physics, [27] affectionately known as the "rubber bible", is an indispensable tool for every scientist, nuclear or otherwise. Although it is not formally considered as an annual publication, a new edition has been published nearly every year since 1914 and it is probably the most comprehensive collection of scientific data in one volume in the world. Another important data book is the *American Institute of Physics Handbook*, [28] which contains long lists of references to original papers. If something more than just the basic data is required, a useful source is Condon's *Handbook of Physics* [29] which can be described as a one-volume synthesis of present day knowledge in the field of physics. Nearly 100 contributors have combined to produce this authoritative account, of which two full sections relate specifically to nuclear science.

ENCYCLOPAEDIAS

The most up-to-date single volume encyclopaedia in the nuclear field is *Newnes Concise Encyclopaedia of Nuclear Energy* [30] in which the entries vary from short definitions to fairly long articles. Each of the signed contributions is the work of a recognized expert and each provides a simply worded but accurate account of a particular concept, so that the book is both useful to the specialist and intelligible to the layman.

The most authoritative encyclopaedia on all branches of physics is the *Handbuch der Physik* [31] published in 54 volumes by Springer-Verlag, although not all of these volumes have yet been issued. Each volume, or group of volumes, covers a specific subject field which is presented in a series of articles by internationally recognized authorities. The number of articles in each volume varies, and in the whole encyclopaedia about 70 per cent of the articles are in English, 25 per cent in German and the rest in French. A German–English and an English–German index is included in each volume. Those of most interest from the viewpoint of atomic energy are listed in the reference at the end of the chapter.

A more modest publication, but entirely in English, is the *Encyclopaedic Dictionary of Physics* [32] in eight volumes, in which a straightforward alphabetical arrangement is used for the entries, the last volume being a subject index. Nuclear science is well-represented, and in fact a high proportion of the editors and contributors come from such places as Harwell, Brookhaven and Oak Ridge. A ninth volume consisting of a six-language glossary is in course of preparation and it is intended to issue annual supplementary volumes which will cover new topics, new developments in topics previously covered and some topics not included in previous volumes for one reason or another.

A less elaborate, but none-the-less useful one-volume encyclopaedia is the *International Dictionary of Physics and Electronics*, [33] which is intended mainly for quick reference. Short entries, arranged alphabetically, are of the encyclopaedic rather than dictionary variety, and are preceded by a brief historical introduction tracing the connection between classical physics and modern atomic and nuclear physics.

DICTIONARIES AND GLOSSARIES

The earliest glossary of nuclear terms to come into general use was that published by the American Society of Mechanical Engineers [34] in 1957 as an A.S.A. standard. In the following year the United Nations issued a short glossary [35] in English with equivalent terms in three other languages. Both these glossaries are now a little out-of-date but have formed a basis for later and more comprehensive publications. A glossary compiled primarily for the non-scientist [36] has been issued by the UKAEA, the fourth edition appearing in 1962. This contains a number of terms relating specifically to British laboratories, installations and experiments which would not normally appear in a general glossary.

The most up-to-date and comprehensive glossary in English is the *Glossary of Terms used in Nuclear Sciences* [37] published by the British Standards Institution in 1962 with the co-operation of a number of interested organizations, including the UKAEA. This standard owes much to the earlier glossaries and its definitions would be acceptable in most English-speaking countries.

Atomic energy is well served by multi-lingual dictionaries, most of them of European origin. *Elsevier's Dictionary of Nuclear Science and Technology* [38] gives short definitions in English of each English term followed by the equivalent terms in French, Spanish, Italian, Dutch and German and indexes in each of these languages are provided. Béné's dictionary [39] was designed primarily for the use of interpreters at international conferences and follows the arrangement of *Elsevier's Dictionary* except that no definitions are given and the languages are confined to English, French, German and Russian. The *Dictionary of Nuclear Physics and Technology* [40] covers the same four languages and gives some indication of variations in the meanings of terms. There is a separate section for each language, with the equivalent terms for the other three on the same page, so that although bulkier, the dictionary is easier to use than the two previously mentioned.

BOOKS

It is difficult to make a realistic estimate of the number of books which have been published on atomic energy even if only those published in English are considered. The two volumes of *British Scientific and Technical Books* [41] list over 100 British books published between 1935 and 1957 and over 200 are listed in the two volumes of the *Cumulated Subject Catalogue* [42] of the British National Bibliography covering the period from 1951 to 1959. The

Harwell Library issue, from time to time, a reading list [43] of British books on atomic energy and the 1964 edition of this contained 237 entries. Hawkins [44] lists nearly 100 American books in the second edition published in 1958 but this list is highly selective and there are nearly 200 listed in the 1963 edition of the USAEC booklet *Technical Books and Monographs Sponsored by the USAEC* which was described in Chapter 3 (reference 5). Apart from these guides reference should be made to publishers' catalogues, particularly those who tend to specialize in nuclear science.

Since it would be impossible in the space available to give adequate guidance on this mass of literature only a few of the more important titles in each category are selected for mention here.

Books for the Non-scientist

The advent of atomic energy has had far reaching effects on the political, social and cultural life of mankind and there is a continuing demand for books which will enable the average person to understand something of the issues which this scientific revolution has raised. In general books of this kind fall into two categories; those that try to explain the basic scientific principles in simple language and those which discuss the social and political implications directly. Typical of books in the first category is Asimov's *Inside the Atom*, [45] which was written for people with no previous knowledge of physics. In a similar vein is Allibone's *Release and Use of Atomic Energy* [46] which includes chapters on nuclear power, radioisotopes and controlled fusion. The second category is typified by Titterton's *Facing the Atomic Future* [47] which is concerned to ensure that citizens of a democracy fully understand the issues involved in the use of nuclear power, nuclear weapons and radiation.

Although atomic energy, as such, is a comparatively new subject field, it is a logical development from the advances made in physics during the last 150 years and the general reader can often obtain a better understanding of the basic ideas of nuclear science by studying this background. Two very readable accounts of modern physics are *The Atom* [48] by Sir G. Thomson, and *Atoms and the Universe*, [49] written by three scientists of international repute. In both these books the material is presented at a slightly more advanced level than in the books previously mentioned.

Between the general reader and the student scientist are several groups of individuals whose interest in atomic energy is such as to warrant special expositions of the subject, tailored to meet their particular requirements. Three typical examples of such an approach are mentioned here. *Modern Physics*, [50] by Caro, McDonnell and Spicer, is an introduction to atomic and nuclear physics for sixth formers who already have some knowledge of elementary physics and mathematics, and is suitable for adults with roughly the same qualifications. *Atomic Energy in Industry*, [51] by Meyer, is designed for tradesmen and technicians, with little or no training in physics, who are working in atomic plant and are particularly concerned with radioactivity and radiological protection. *Modern Nuclear Technology*, [52] which has a number of contributors, is based on a series of lectures given at the University of California as part of a course entitled, "A survey of nuclear engineering for management", the purpose being to present the fundamentals of the subject at a level suitable for managers and administrators with some engineering background.

BOOKS FOR SCIENTIFIC AND TECHNICAL READERS

There are a number of introductions to the subject for readers with scientific training, of which the best known is Glasstone's *Sourcebook on Atomic Energy,* [53] which is sponsored by the USAEC, and is a compilation of information on all aspects of atomic energy, suitable for a wide variety of technical readers. The author uses an historical approach, giving full details of individual discoveries but very few references to original papers. A more formal textbook is Glassner's *Introduction to Nuclear Science* [54] which grew out of an intensive course for high school science teachers. A similar approach is used in *Nucleonics Fundamentals* [55] based on courses given to naval officers at the General Line and Naval Science School of the U.S. Navy.

The large volume of scientific and technical information generated by the war-time "Manhattan Project", which led to the making of the first bomb, has been published systematically in book form in the fifty volumes of the *National Nuclear Energy Series,* full details of which are given in the USAEC publication *Technical Books and Monographs* which has already been referred to. The last volume in this series appeared in 1955. In the same year the USAEC compiled their *Selected Reference Material on Atomic Energy* [56] in eight volumes, sets of which were presented to the official atomic energy projects of those countries which were represented at the first International Conference on the Peaceful Uses of Atomic Energy, Geneva, 1955. These volumes were subsequently published by the U.S. Government Printing Office in paperbacked editions. Between them they constitute a detailed scientific account of the state of atomic energy research and development in America at that time. In 1958, at the time of the second Geneva conference, the USAEC again issued a presentation set of twelve volumes in the *Atom for Peace Series,* [57] which were subsequently issued separately by Addison-Wesley.

The proceedings of the two Geneva conferences [58, 59] are themselves sources of hundreds of original papers on all aspects of nuclear energy and are still extensively quoted. The information made available at Geneva has provided adequate material for a number of publications such as Van Nostrand's *Geneva Series on the Peaceful Uses of Atomic Energy* and Pergamon's *Progress in Nuclear Energy* in twelve series, of which the earlier volumes in each series drew heavily on the Geneva papers. Further information on these series can be found in the publishers' catalogues.

For many people the origins of nuclear science are inextricably linked with the atomic bomb, and the war time effort to develop this weapon. By any standards the six years required to translate a tentative theory into an accomplished fact was fantastically short, a rate of development almost without precedent in the annals of science. The story of this period is told in an officially sponsored account [60] by H. D. Smyth, and in an unofficial but, in some ways, more revealing account by W. L. Laurence [61] who was the only newspaper reporter present at the first test explosion of the bomb in New Mexico in July 1945, and later prepared the official press releases on the bombing of Hiroshima and Nagasaki.

This eventful period in American history is described rather more fully in the first volume of the *History of the USAEC* [62] covering the years from 1939 to 1946. Written from the perspective of sixteen years later it reveals a good deal more information, particularly at the political and administrative level, than will be found in the earlier books.

Although the bomb was born in America it was conceived in Europe and to obtain a complete picture of the period some account must be taken of the part played by Britain in the events leading up to the development of the bomb. This has been described in detail by the UKAEA's official historian in *Britain and Atomic Energy, 1939–1945*, [63] which is an authoritative account based on official records and on interviews with many of the people involved.

CONFERENCE PROCEEDINGS

Librarians frequently receive requests for proceedings of conferences and symposia, or for individual papers presented at such meetings, and this information is often difficult to identify and even more difficult to locate. It is well known that of all the material included in abstract journals, conference literature is probably the most inadequately represented and recourse to other sources is usually necessary. In the atomic energy field the problem is not quite so severe because there are a number of excellent guides to conference literature.

In many cases the first task is to identify the name, date and sponsoring body of a particular conference and in the nuclear field the most helpful publication for this is the IAEA's *Conferences, Meetings, Training Courses in Atomic Energy*, [64] which appears bimonthly and covers national and international conferences which are of interest to nuclear scientists and technologists. This publication also includes, in each issue, a list of forthcoming IAEA conferences and a bibliography of published proceedings of past IAEA meetings.

A number of nuclear periodicals contain lists of forthcoming conferences, although the emphasis is usually on the meetings held in the country in which the journal is published. An exception is the *Calendrier International des Conferences, Congres, Expositions*, [65] which is issued by the French Atomic Energy Commission as a supplement to the *Bulletin d'Informations Scienti-*

fiques et Techniques. This covers national and international meetings in many countries and includes conferences on astronautics and space research as well as atomic energy.

Covering a wider subject field, but listing only international scientific conferences, is Part 1 of the *World List of Future International Meetings,* [66] issued monthly by the Library of Congress. The March, June, September and December issues list all international meetings planned for the ensuing three years about which information is available. The intervening issues list only new meetings and changes in previously listed meetings.

A similar publication is *Forthcoming International Scientific and Technical Conferences,* [67] published by the Department of Scientific and Industrial Research in London, which also includes annual general meetings of professional institutions.

Many of the smaller and more informal meetings which are held from time to time do not appear in the publications described above and many documentation centres maintain their own information files on these conferences. Such a file has been built up over a number of years by the USAEC Division of Technical Information at Oak Ridge and this information is now being made generally available in a periodically issued list called *Availability of Nuclear Science Conference Literature* [68] which, in spite of its title, is not a bibliography of conference proceedings or papers but a listing of those meetings of interest to the users of the USAEC Information Services. The inclusion of a conference in this list indicates that DTIE will try to obtain information about the availability of proceedings or papers, to acquire such proceedings or papers and to include in future listings any information which is likely to be helpful to users. Whereas the lists published by the IAEA, Library of Congress and DSIR are intended primarily to notify scientists of future meetings, the USAEC lists are issued as a tool for librarians and documentalists and are oriented in their direction. They show where proceedings have been published or are to be published, if this is known, and indicate when individual papers have been received at Oak Ridge, but they do not itemize them.

Assuming that a conference or meeting has been identified, the next problem is to obtain detailed information about the proceedings or about the individual papers. In the nuclear field the most useful sources are *N.S.A.* and the Gmelin *Indexed Bibliography.* Published proceedings have been abstracted in *N.S.A.* for many years and are indexed in the subject indexes under the heading "Conferences". A computer-compiled index to all the conference entries in volumes 11 to 16 (1957–1962) of *N.S.A.* has been issued as a separate report [69] and this will be kept up-to-date by the publication of revisions. This is an exceptionally useful document since it provides access to conference information under (i) permuted title and place held; (ii) date of conference; (iii) number, in the case of numbered conferences (e.g. Second International Conference on the Peaceful Uses of Atomic Energy); and (iv) report number, if proceedings were published in this form. Later revisions will also serve as an index to the *Availability of Nuclear Science Conference Literature* (TID–19000) already mentioned.

From volume 17, number 18, September 30, 1963, *N.S.A.* has also listed individual papers received prior to publication of the full proceedings. These are identified under a CONF number (CONF–1, CONF–2, etc.) which is the same for all papers presented at a particular conference, and are indexed under the CONF number in the report number indexes. When conference proceedings are later published the availability of the proceedings as a whole is shown and entries for individual papers dropped. Individual conference papers listed in *N.S.A.* are acquired by the AEC in two ways. They are either received direct at Oak Ridge or they are obtained by the Gmelin Institute's New York Office as a result of Gmelin's positive programme for acquiring all conference papers in the nuclear field, and copies forwarded to Oak Ridge. In most cases single copies only are obtained by either method. The actual availability of these papers, therefore, varies considerably. Some can be obtained on inter-library loan from the Oak Ridge Institute of Nuclear Studies (see Chapter 3); others can be obtained only from Gmelin. All conference papers acquired by the Gmelin Centre for Atomic Energy Documentation are listed in the AED *Indexed Bibliography*, series A–B together with complete conference proceedings. In most cases AED are prepared to supply photocopies or microcopies of these papers at nominal charges.

The USAEC itself has sponsored many conferences, symposia and meetings and the proceedings of such meetings have frequently been published in report form. *Proceedings of Technical Meetings*, [70] issued by the AEC in 1961, contains annotated entries for all conferences for which the AEC has been sponsor, co-sponsor or a major participant during the period 1946 to 1961. Entries are grouped under broad subject headings and subject and report number indexes are included.

TRANSLATIONS

English-speaking peoples are not renowned for their ability to communicate in other languages and their scientists are no exception to this rule. A proportion may have sufficient knowledge of French or German to be able to extract the gist of an article in one of these languages and at one time this ability was probably all that was required. The situation has, however, changed dramatically since the Soviet Union became a first-rank scientific nation and the volume of scientific and technical information now appearing in the Russian language is second only to that appearing in English.

Until Russian becomes a normal part of the school curriculum in English-speaking countries, the only practical solution to this problem is a large-scale translation programme, especially in subject fields such as atomic energy where Soviet technology is particularly advanced. Since the number of competent scientific Russian translators in western countries is somewhat limited, it has been necessary to seek solutions on a national and even international level, in order to make the best use of the talent available.

Basically two approaches have been made to this problem. The first is that of providing cover-to-cover translations of Russian periodicals and other publications and making these commercially available. The second is that of setting up a number of translations pools which collect or index translations of individual articles and reports and publish periodical listings of these.

Cover-to-cover Translations

The growth of cover-to-cover translations of Russian periodicals since 1950 has been phenomenal. In addition to these a number of Soviet journals are translated in part only, the criteria for selection varying in each case. Most of the important Russian periodicals covering atomic energy fall into one or other of these categories, the chief agencies being the American Institute of Physics, [71] Consultants Bureau, [72] Columbia Technical Translations, [73] and the Office of Technical Services [74] in America, and the Department of Scientific and Industrial Research [75] in Britain. Detailed information on translated Russian scientific journals will be found in the *List of Russian Scientific Journals available in English* [76] issued by the National Science Foundation in 1961. A similar list has been published by the National Lending Library in Britain (see Chapter 4, reference 6). More up-to-date lists appear from time to time in *Technical Translations* [78] and the *NLL Translations Bulletin.* [79]

Translations Pools

In the United States the two main centres, covering all fields of science and technology, and all languages, are the Office of Technical Services in Washington and the Special Libraries Association Translations Center in the John Crerar Library, Chicago. The two centres work in close co-operation and can provide substantially the same information. OTS collects translations from Government agencies in the U.S.A. and abroad, whereas the SLA Center concentrates on translations from domestic and foreign non-Government organizations; copies of translations are then exchanged. From 1953 to 1958 the SLA Center issued lists of its holdings [77] under various titles, but these were discontinued when OTS undertook to list the SLA translations in *Technical Translations* [78] which is probably the most comprehensive list of translation now available and covers translations in progress as well as those completed. OTS can either supply, at a nominal cost, copies of the translations listed or indicate where they may be obtained.

The main translations centre in Britain is the National Lending Library for Science and Technology which concentrates mainly on translations from the Russian and exchanges translations and information about them with OTS. Translations acquired are listed in the *NLL Translations Bulletin.* [79] The British file of the *Commonwealth Index of Unpublished Scientific and Tech-*

nical Translations is held by Aslib, 3 Belgrave Square, London S.W.1. This is not a translation pool but a card file of about 150,000 references to unpublished translations reported by a number of co-operating centres in the Commonwealth, and includes references to articles in cover-to-cover translated journals.

In Canada the principal centre is the National Research Council Library in Ottawa which publishes, periodically, a *List of Technical Translation* and also maintains the Canadian file of the Commonwealth Index.

The only translations pool devoted exclusively to atomic energy is the Euratom Transatom Service in Brussels, a centre created jointly by Euratom, the USAEC and the UKAEA in 1960 to pool their efforts in collecting and distributing information on translations of atomic energy literature, mainly from the Russian and other Eastern European languages. The Service publishes *Transatom Bulletin* [80] which gives details of translations completed and in progress, with the name of the issuing agency. Each issue of the *Bulletin* also contains a list of all cover-to-cover translations of Russian journals of interest in the atomic energy field.

OTHER AGENCIES

The USAEC has sponsored a number of translations of books, journal articles, conference proceedings and reports, the majority of which are issued in the AEC-tr series and are announced and indexed in *N.S.A.*, distributed to official recipients and depository libraries and sold by the Office of Technical Services. A catalogue [81] of the most important of these translations has been issued by the Division of Technical Information. All translations in progress and completed are notified to OTS and to Euratom for inclusion in *Technical Translations* and *Transatom Bulletin* respectively.

A cumulative listing of translations held by DTI has been published as *Translation Title List and Cross Reference Guide* [82] with a separate subject index.

Translations made by the UKAEA are listed in *UKAEA List of Publications available to the Public* (see Chapter 2, reference 15), and many of them are distributed to official atomic energy projects and to depository libraries. In certain cases only a few copies of translations are made and these can be borrowed from the originating library and in some instances from the NLL as well. Notifications of UKAEA translations are sent to the Aslib Commonwealth Index and to Euratom for listing in *Transatom Bulletin*.

The Organization for Economic Co-operation and Development (OECD), as part of its policy of encouraging the use of Russian and Eastern European scientific literature, has set up a European Translation Centre (ETC) at Delft in the Netherlands, the aim of which is to collect unpublished translations, or information about them, and to make copies available at a nominal cost. The Centre operates mainly through national centres and translations acquired are listed in the OTS publication *Technical Translations*. Information about the work of ETC and a list of the national centres will be found in each issue of *Russian Technical Literature* [83] published in both English and French editions by OECD. This journal also gives details of Russian books translated or in process of translation by commercial publishers, and reviews new language dictionaries and glossaries.

More detailed information on the network of organizations involved in the translation of Russian scientific literature into English will be found in a booklet [84] compiled by Massachusetts Institute of Technology and published by the National Science Foundation in 1962, which gives information on the procurement and availability of Russian literature as well as its translation.

REFERENCES

Bibliographies and Abstracts

1. UNITED NATIONS ATOMIC ENERGY COMMISSION. *An international bibliography on atomic energy.* United Nations, New York, 1949—1954.
   Vol. 1, Political, economic and social aspects, March 1949 (Supplement 1, July 1950. Supplement 2, September 1953); Vol. 2, Scientific aspects, May 1951 (Supplement 1, May 1952. Supplement 2, March 1954).
2. UNITED STATES ATOMIC ENERGY COMMISSION. *Selected readings on atomic energy.* U.S. Government Printing Office, Washington, D.C., 1958.
3. ANTHONY, L. J. *Books on atomic energy. Part 1, Nuclear power. Part 2, Radioisotopes and radiations.* British Book News, nos. 212 and 213, April and May, 1958.
4. *Nuclear Science Abstracts.* (Semi-monthly.) Published by the Division of Technical Information Extension, USAEC, and sold by U.S. Government Printing Office, Washington 25, D.C. 1947 and onwards.
   Each issue has corporate author, personal author, subject and report number indexes and these are cumulated quarterly, half-yearly and annually. Cumulative indexes covering longer periods are available as follows:
   Vols. 1—4 (1947—1950), Cumulated author and subject indexes are included in vol. 4, available from the Johnson Reprint Corporation, 111 Fifth Avenue, New York; Vols. 5—10 (1951—1956), Cumulated author and subject indexes available from USGPO; Vols. 11—15 (1957—1961), Cumulated personal author, corporate author and subject indexes available from USGPO.
5. *Science Abstracts.* (Monthly.) Institution of Electrical Engineers, Savoy Place, London W.C.2. 1898 and onwards.
   Section A—Physics abstracts; Section B—Electrical engineering abstracts.
6. *Chemical Abstracts.* (Semi-monthly.) American Chemical Society, 20th and Northampton Streets, Easton, Pennsylvania, 1907 and onwards.
7. *U.S. Government Research Reports.* (Semi-monthly.) Office of Technical Services. U.S. Department of Commerce, Washington D.C. 1946 and onwards.
8. *Nuclear Engineering Abstracts.* (Irregular.) Silver End Documentary Publications Ltd., 9—11 Tottenham St., London, W.1. 1960 and onwards.

The publication programme as at June 1964 was as follows:

Vol. no.	Period covered	Scheduled publication date	Date actually published
Vol. 1, no. 1	January—March 1960	July 1960	July 1960
Vol. 1, no. 2	April—June 1960	October 1960	April 1961
Vol. 2, no. 1	July—September 1960	—	January 1962
Vol. 2, no. 2	October—December 1960	—	December 1962
Vol. 3	January—December 1961	Not yet published	
Vol. 4, no. 1	January—April 1962	January 1963	October 1963
Vol. 4, no. 2	May—August 1962	May 1963	June 1964
Vol. 4, no. 3	September—December 1962	September 1963	Not yet published

9. *Engineering Index.* (Monthly.) Engineering Index Inc., 345 East 47th St., New York 17. 1885 and onwards (annual volume only before 1962).
10. *Bulletin Signalétique. Section 5, Physique Nucléaire.* (Monthly.) Centre de Documentation du Centre National de la Recherche Scientifique, 15 Quai Anatole-France, Paris 7e. 1940 and onwards.
11. *Zentralblatt für Kernforschung und Kerntechnik.* (Monthly.) Institut für Dokumentation der Deutschen Akademie der Wissenschaften zu Berlin. Published by Akademie-Verlag G.m.b.H., Leipziger Straße 3—4, 108 Berlin, 1961 and onwards.
12. *Atomkernenergie Dokumentation, Indexed Bibliography, series A—B.* (Monthly.) Zentralstelle für Atomkernenergie Dokumentation beim Gmelin-Institut, Carl-Bosch-Haus, Varrentrappstrasse 40—42, Frankfurt (Main), Germany, 1962 and onwards.
    Series A, published separately December 1957 to 1961, Series B, published separately July 1958 to 1961.
13. *Atomkernenergie Dokumentation, series C, Bibliographic Review of Selected Subjects.* (Irregular.) Zentralstelle für Atomkernenergie Dokumentation beim Gmelin-Institut, Frankfurt (Main), Germany, 1959 and onwards.
14. NATIONAL FEDERATION OF SCIENCE ABSTRACTING AND INDEXING SERVICES. *A guide to the world's abstracting and indexing services in science and technology.* Report no. 102. The Federation, 374 East Capitol St., Washington 3, D.C., 1963.

Directories and Handbooks

15. HASLETT, A. W., ED. *World nuclear directory: an international reference book.* 2nd ed., enl. and rev. London, Harrap, 1963. 626 p.
16. ALMAN, M., ED. *Aslib directory: a guide to sources of information in Great Britain and Ireland.* Aslib, 3 Belgrave Sq., London, S.W.1. 1957. 2 vols.
    Amendment list issued in 1961 covering changes occurring during 1957—1960.
17. *Specialised Science Information Services in the United States.* National Science Foundation, Washington 25, D.C., November 1961.
    The section on Nuclear Physics and Technology lists 13 agencies and that on Physics 19 agencies.
18. *Directory of Selected Scientific Institutions in the U.S.S.R.; with an introduction to the administration of science and technology in the U.S.S.R.* National Science Foundation, Washington 25, D.C., 1963.

19. HASLETT, A. W., ED. *Who's who in atoms*. 3rd ed. London, Harrap, 1962. 1186 p.
20. CATTELL, J., ED. *American men of science; a biographical directory*. 10th ed. Jaques Cattell Press Inc., Arizona State University, Tempe, Arizona, 1960.
    Vols. 1–4, Physical and biological sciences; Vol. 5, Social and behavioural sciences.
21. *Directory of British Scientists, 1964–1965*. London, Ernest Benn, 1964. 2001 p.
    Contains an index to names, grouped under broad subject headings and a list of scientific and technical journals.
22. HOGERTON, J. F. AND OTHERS. *The atomic energy deskbook; prepared under the auspices of the Division of Technical Information, USAEC*. New York, Reinhold; London, Chapman and Hall, 1963. 673 p.
    Includes a list of nuclear periodicals.
23. ETHERINGTON, H., ED. *Nuclear engineering handbook*. New York, McGraw-Hill, 1958.
24. FRISCH, O. R., ED. *The nuclear handbook*. London, Newnes, 1958.
25. JEFFERSON, S., ED. *Handbook of the atomic energy industry*. London, Newnes, 1958.
26. INSTITUT NATIONAL DES SCIENCES ET TECHNIQUES NUCLÉAIRES. *Génie atomique*. (In French.) Presses Universitaires de France, 108 boulevard Saint-Germain, Paris 6e, 1960.
    Vol. 1, General physics, reactor physics, protection against radiation; Vol. 2, Reactor instrumentation and control; Vol. 3, Nuclear reactors; Vol. 4, Nuclear materials, isotopes separation, radioisotopes.
27. HODGMAN, C. D. AND OTHERS, EDS. *Handbook of chemistry and physics: a ready reference book of chemical and physical data*. Chemical Rubber Publishing Co., 2310 Superior Ave. N.E., Cleveland, Ohio. 45th ed., 1963. 3604 p.
28. AMERICAN INSTITUTE OF PHYSICS. *American Institute of Physics handbook*. 2nd ed. New York, McGraw-Hill, 1963.
    Sections relating to atomic energy are: Section 7, Atomic and molecular physics. 229 p. Section 8, Nuclear physics. 327 p.
29. CONDON, E. U. AND ODISHAW, H., EDS. *Handbook of physics*. New York, McGraw-Hill, 1958.
    Sections of interest are: Part 7, Atomic physics. 10 chapters. 173 p. Part 9, Nuclear physics. 13 chapters. 257 p.
    A list of references and a bibliography is given at the end of each chapter.

Encyclopaedias

30. BARNES, D. E. AND OTHERS, EDS. *Newnes concise encyclopaedia of nuclear energy*. London, Newnes, 1962. 886 p.
31. FLUGGE, S., ED. *Handbuch der Physik (Encyclopaedia of physics)*. Berlin, Springer-Verlag, 1955 and onwards. 54 vols.
    Volumes concerned with atomic energy are:
    Group VI: X-rays and corpuscular rays. Vols. 30–34
        Vol. 30   X-rays. 1957. 384 p.
        Vol. 31   Corpuscles and radiation in matter, I. (Not yet issued.)
        Vol. 32   Structural research. 1957. 663 p.
        Vol. 33   Optics of corpuscles. 1956. 702 p.
        Vol. 34   Corpuscles and radiation in matter, II. 1958. 316 p.

    Group VII: Atomic and molecular physics. Vols. 35–37
        Vol. 35   Atoms, I. 1957. 454 p.
        Vol. 36   Atoms, II. 1956. 424 p.
        Vol. 37   Part 1. Atoms, III; Molecules, I. 1959. 439 p.
        Vol. 37   Part 2. Molecules II. 1961. 303 p.

Group VIII: Nuclear physics. Vols. 38—45

    Vol. 38   Part 1. External properties of atomic nuclei. 1958. 471 p.
    Vol. 38   Part 2. Neutrons and related gamma ray problems. 1959. 868 p.
    Vol. 39   Structure of atomic nuclei. 1957. 566 p.
    Vol. 40   Nuclear reactions, I. 1957. 553 p.
    Vol. 41   Part 1. Nuclear reactions, II. 1959. 580 p.
    Vol. 41   Part 2. Beta decay. 1962. 117 p.
    Vol. 42   Nuclear reactions, III. 1957. 626 p.
    Vol. 43   Mesons. (Not yet issued.)
    Vol. 44   Nuclear instrumentation, I. 1959. 473 p.
    Vol. 45   Nuclear instrumentation, II. 1958. 544 p.
    Vol. 46   Part 1. Cosmic rays, I. 1961. 333 p.
    Vol. 46   Part 2. Cosmic rays, II. (Not yet issued.)

32. THEWLIS, J. AND OTHERS, EDS. *Encyclopaedic dictionary of physics.* Oxford, Pergamon Press, 1961—1963. 8 vols.

33. MICHELS, W. C. AND OTHERS, EDS. *International dictionary of physics and electronics.* 2nd ed. New York, London, Van Nostrand, 1961.
With French—English, German—English, Russian—English and Spanish—English indexes.

Dictionaries and Glossaries

34. NATIONAL RESEARCH COUNCIL. CONFERENCE ON THE GLOSSARY OF TERMS IN NUCLEAR SCIENCE AND TECHNOLOGY. *A glossary of terms in nuclear science and technology.* American Society of Mechanical Engineers, 29 West 39th St., New York, 1957. 188 p. (ASA N 1. 1—1957).

35. UNITED NATIONS. *Atomic energy: glossary of technical terms.* United Nations, New York, 1958.
A list of terms in English with equivalents in French, Spanish and Russian and indexes in the last three languages.

36. UNITED KINGDOM ATOMIC ENERGY AUTHORITY. *Glossary of atomic terms.* 4th ed. London, H.M.S.O., 1962. 60 p.

37. BRITISH STANDARDS INSTITUTION. *Glossary of terms used in nuclear science.* B.S.I., 2 Park St., London, W.1, 1962. (B.S. 3455: 1962.)

38. CLASON, W. E., ED. *Elsevier's dictionary of nuclear science and technology in six languages.* Amsterdam, New York, London, Elsevier, 1958. 914 p.
English entries are serially numbered and indexes in other languages refer to these numbers. A supplementary index for Russian was published in 1961.

39. BENE, G. J. AND OTHERS. *Nuclear physics and atomic energy: terms of nuclear physics and nuclear technology in English, French, German and Russian.* Amsterdam, New York, London, Elsevier, 1960.

40. SUBE, R. *Dictionary of nuclear physics and technology.* Oxford, Pergamon Press, 1962. 1600 p.

Books

41. (a) ASLIB. *British scientific and technical books: a select list of recommended books published in Great Britain and the Commonwealth in the years 1935 to 1952.* London, J. Clarke & Co. 1956. 364 p.
(b) ANTHONY, L. J., ED. *British scientific and technical books 1953—1957.* London, J. Clarke & Co. (for Aslib), 1960. 251 p.

42. (a) *British National Bibliography Cumulated Subject Catalogue 1951—1954.* Council of the British National Bibliography, Ltd., British Museum, London, W.C.1, 1958.

(b) *British National Bibliography Cumulated Subject Catalogue 1955—1959.* Council of the B.N.B., 1963.
43. ATOMIC ENERGY RESEARCH ESTABLISHMENT, HARWELL. *A list of British books and periodicals on atomic energy: compiled and revised by N. B. Skeats.* Information Office, AERE, Harwell, Berks. July 1964. AERE Reading List 3 (2nd revision).
44. HAWKINS, R. R. *Scientific, medical and technical books published in the USA: a selected list of titles in print with annotations.* 2nd ed. Books published to December 1956. National Research Council, Washington, 1958. 1491 p.
45. ASIMOV, I. *Inside the atom.* London, Abelard-Schuman Ltd., 1956. 176 p.
46. ALLIBONE, T. E. *The release and use of atomic energy.* London, Chapman & Hall, 1961. 170 p.
47. TITTERTON, E. W. *Facing the atomic future.* London, Macmillan, 1956.
48. THOMSON, SIR G. *The atom.* 6th ed. London, O.U.P., 1962. 232 p.
49. JONES, G. O., ROTBLAT, J. AND WHITROW, G. J. *Atoms and the universe: an account of modern views of the structure of matter and the universe, with a prefatory note by Sir John Cockroft.* 2nd ed. London, Eyre & Spottiswoode, 1962. 276 p.
50. CARO, D. E., MCDONNELL, J. A. AND SPICER, B. M. *Modern physics: an introduction to atomic and nuclear physics.* London, E. Arnold, 1962. 222 p.
51. MEYER, L. *Atomic energy in industry: a guide for tradesmen and technicians.* Chicago, American Technical Society, 1963. 128 p.
52. MILLS, M. M., BIEHL, A. T. AND MAINHARDT, R., EDS. *Modern nuclear technology: a survey for industry and business.* New York, McGraw-Hill, 1960. 336 p.
53. GLASSTONE, S. *Sourcebook on atomic energy.* 2nd ed. New York, London, Van Nostrand, 1958. 614 p.
54. GLASSNER, A. *Introduction to nuclear science.* New York, London, Van Nostrand, 1961. 213 p.
55. HOISINGTON, D. B. *Nucleonics fundamentals.* New York, McGraw-Hill, 1959. 410 p.
56. UNITED STATES ATOMIC ENERGY COMMISSION. *Selected reference materials on atomic energy.* U.S. Government Printing Office, Washington D.C., 1955. 8 vols. Vol. 1: Research reactors; Vol. 2: Reactor handbook, part 1, physics; Vol. 3: Reactor handbook, part 2, engineering; Vol. 4: Reactor handbook, part 3, materials; Vol. 5: Neutron cross sections; Vol. 6: Chemical processing and equipment; Vol. 7: Eight-year isotope summary; Vol. 8: Information sources.
57. UNITED STATES ATOMIC ENERGY COMMISSION. *Atoms for peace series.* Reading, Massachusetts, Addison-Wesley Pub. Co., 1958. 12 vols.
Subsequently published as separate volumes as follows:
Bishop, A. A. Project Sherwood—the U.S. programme in controlled fusion. 288 p.
Chastain, J. W., ed. U.S. research reactor operation and use. 384 p.
Claus, W. D., ed. Radiation biology and medicine. 968 p.
Clegg, J. W. and Foley, D. D. eds. Uranium ore processing. 450 p.
Cuthbert, F. L. Thorium production technology. 320 p.
Dietrich, J. R. and Zinn, W. H. Solid fuel reactors. 864 p.
Holden, A. N. Physical metallurgy of uranium. 272 p.
Kramer, A. W. Boiling water reactors. 592 p.
Lane, J. A. and others, eds. Fluid fuel reactors. 1008 p.
Seaborg, G. T. The transuranium elements. 350 p.
Starr, C. and Dickinson, R. W. Sodium graphite reactors. 304 p.
USAEC Naval Reactors Branch. The Shippingport pressurised water reactor. 600 p.
58. INTERNATIONAL CONFERENCE ON THE PEACEFUL USES OF ATOMIC ENERGY, GENEVA, AUGUST, 1955. *Proceedings.* United Nations, 1956. 16 vols.

59. INTERNATIONAL CONFERENCE ON THE PEACEFUL USES OF ATOMIC ENERGY, GENEVA, SEPTEMBER, 1958. *Proceedings.* United Nations, 1959. 33 vols.
Individual volumes of both these series are listed in *H.M.S.O. Sectional List no. 63, Atomic Energy* (see Chapter 2, reference 19).
60. SMYTH, H. D. *A general account of the development of methods of using atomic energy for military purposes under the auspices of the United States Government, 1940—1945.* Washington, U.S. Government Printing Office; London, H.M.S.O. 1945.
61. LAURENCE, W. L. *Dawn over zero: the story of the atomic bomb.* 2nd ed. New York, A. A. Knopf, 1947, 289 p.
62. HEWLETT, R. G. AND ANDERSON, O. E. *The New World, 1939—1946; volume 1 of a history of the USAEC.* Pennsylvania State University Press, 1962. 766 p.
63. GOWING, M. *Britain and atomic energy, 1939—1945.* London, Macmillan; New York, St. Martin's Press. 1964.

Conference literature

64. INTERNATIONAL ATOMIC ENERGY AGENCY. *Conferences, meetings, training courses in atomic energy.* (Bi-monthly.) IAEA, Kärntnerring 11, Vienna 1. 1959 and onwards.
Includes a list of periodicals and other sources from which the information about conferences has been obtained.
65. *Calendrier International des Conférences, Congrès, Expositions.* (Irregular.) Supplement to Bulletin d'Informations Scientifique et Techniques, Editions Dunod, 92 rue Bonaparte, Paris 6e, 1957 and onwards.
66. *World List of Future International Meetings. Part 1. Science, Technology, Agriculture, Medicine.* (Monthly.) U.S. Government Printing Office, Washington 25, D.C.
Compiled by the International Organizations Section, General Reference and Bibliography Division, Library of Congress. Each issue has a subject index, place name index and index to organizations sponsoring meetings.
67. DEPARTMENT OF SCIENTIFIC AND INDUSTRIAL RESEARCH. *Forthcoming international scientific and technical conferences.* DSIR, Overseas Liaison Group, Africa House, Kingsway, London, W.C.2.
Issued half-yearly with quarterly supplements.
68. UNITED STATES ATOMIC ENERGY COMMISSION. *Availability of nuclear science conference literature.* (TID—19000.) USAEC, Division of Technical Information Extension, P.O. Box 62, Oak Ridge, Tennessee. List no. 1, July 1963 and onwards. (Irregular.)
69. UNITED STATES ATOMIC ENERGY COMMISSION. *Index to conferences relating to nuclear science.* (TID—4043—Rev. 1.) USAEC, DTIE, P.O. Box 62, Oak Ridge, Tennessee, May 1964.
70. UNITED STATES ATOMIC ENERGY COMMISSION. *Proceedings of technical meetings* (1946—1961). USAEC, DTIE, P.O. Box 62, Oak Ridge, Tennessee, October 1961.

Translations

71. *Cover-to-cover translations of Russian journals published by the American Institute of Physics, 335 East 45th St., New York 17.*
Doklady Akademii Nauk SSSR (physics papers only): (Soviet Physics—Doklady, 1956 and onwards); Fizika tverdogo tela: (Soviet Physics—Solid state, 1959 and onwards); Kristallografiya: (Soviet Physics—Crystallography, 1957 and onwards); Uspekhi fizicheskikh nauk: (Soviet Physics—Uspekhi, 1958 and onwards); Zhurnal eksperimental'noi i teoreticheskoi fiziki: (Soviet Physics—JETP, 1955 and onwards); Zhurnal tekhnicheskoi fiziki: (Soviet Physics—Technical physics, 1956 and onwards).

72. *Cover-to-cover translations published by Consultants Bureau Enterprises Inc., 227 West 17th St., New York 11.*
    Atomniya energiya: (Soviet journal of atomic energy, 1956 and onwards); Doklady Akademii Nauk SSSR (physical chemistry papers only): (Proceedings of the Academy of Sciences of the USSR, Physical Chemistry Section, 1957 and onwards); Izvestiya Akademii Nauk SSSR, Otdelenie khimicheskikh nauk: (Bulletin of the Academy of Sciences of USSR, Division of Chemical Sciences, 1952 and onwards); Zhurnal analiticheskoi khimii: (Journal of analytical chemistry of the USSR, 1952 and onwards); Zhurnal prikladnoi khimii: (Journal of applied chemistry of the USSR, 1950 and onwards).
73. *Cover-to-cover translations published by Columbia Technical Translations, 5 Vermont Avenue, White Plains, N.Y.*
    Izvestiya Akademii Nauk SSSR, seriya fizicheskaya: (Bulletin of the Academy of Sciences of the USSR; physics series, 1954 and onwards).
74. *Cover-to-cover translations published by Office of Technical Services.*
    Radiokhimiya: (Radiochemistry, 1960 and onwards); Uspekhi fizicheskikh nauk: (Advances in the physical sciences, 1957 only); Zhurnal neorganicheskoi khimii: (Journal of inorganic chemistry, 1957 only).
75. *Cover-to-cover translations sponsored by DSIR.*
    Priborostroenie: (Instrument construction, 1959 and onwards. Published by Taylor & Francis, Red Lion Court, London E.C.4); Uspekhi khimii: (Russian chemical reviews, 1960 and onwards); Zhurnal fizicheskoi khimii: (Russian journal of physical chemistry, 1959 and onwards); Zhurnal neorganicheskoi khimii: (Russian journal of inorganic chemistry, 1959 and onwards).
    The last three are compiled by the Chemical Society, London and published by Cleaver-Hume Press Ltd., 31 Wright's Lane, London W.8.
76. NATIONAL SCIENCE FOUNDATION. *List of Russian scientific journals available in English.* Office of Science Information Service, National Science Foundation, Washington 25, D.C., 1961.
77. (a) SPECIAL LIBRARIES ASSOCIATION TRANSLATION COMMITTEE. *SLA List of translations (covering the period up to 1953).* October 1953.
    (b) JOHN CRERAR LIBRARY. *SLA List of translations: supplement 1953—1954.* September 1954.
    (c) LIBRARY OF CONGRESS, SCIENTIFIC TRANSLATIONS CENTRE. *Bibliography of translations of Russian scientific and technical literature.* Numbers 1—39. October 1953—December 1956.
    (d) *Translations monthly.* Vols 1—4, January 1955 to December 1958.
    All the above are published by SLA Translations Centre, John Crerar Library, 86 East Randolph St., Chicago 1, Illinois.
78. *Technical Translations.* (Semi-monthly.) Office of Technical Services, Washington 25, D.C. 1959 and onwards.
    Each issue has indexes under author, subject, journal of original article, and serial number where these are given. Indexes are cumulated semi-annually.
79. *NLL Translations Bulletin.* (Monthly.) National Lending Library for Science and Technology, Boston Spa, Yorkshire, March 1961 and onwards.
    Continuation of LLU Translations Bulletin (1959—1961) which superseded the DSIR Translated Contents List of Russian Periodicals.
80. *Transatom Bulletin.* (Monthly.) Elsevier Publishing Co., Jan van Galenstraat 335, Amsterdam, Netherlands, 1961 and onwards. (In English.)
    Each issue has author index and index by source journal, and these are cumulated annually.
81. UNITED STATES ATOMIC ENERGY COMMISSION. *Selected technical translations.* 2nd ed. USAEC, DTIE, P.O. Box 62, Oak Ridge, Tennessee, January 1964.

82. UNITED STATES ATOMIC ENERGY COMMISSION. *Translation title list and cross reference guide.* (TID—4025, 1st rev., part 1.) Subject index (TID—4025, 1st rev., part 2.) Office of Technical Services, Washington 25, D.C.
83. *Russian Technical Literature.* (Quarterly.) OECD, Directorate for Scientific Affairs, 2 rue André-Pascal, Paris 16e, 1960 and onwards. (From January 1964 title changed to "Science East to West".)
84. GOROKHOFF, B. I. *Providing U.S. scientists with Soviet scientific information.* Rev. ed. National Science Foundation, Washington 25, D.C., April 1962.

APPENDIX
PERIODICALS AND SERIALS OF GENERAL INTEREST

The following is a select list of periodicals which are of interest in the atomic energy field. These are additional to the journals issued by atomic energy organizations which have been mentioned in previous chapters.

1. *Annual Review of Nuclear Science.* Annual Reviews Inc., 231 Grant Ave., Palo Alto, California, 1952 and onwards. (Annual.)
   Long review articles on nuclear physics and chemistry, particle and high energy physics, radiochemistry, radiation chemistry and radiobiology.
2. *Applied Atomics: world report to business and industry.* Comtelburo Ltd., 85 Fleet Street, London, E.C.4. October 1955 and onwards. (Weekly.)
   News of political, economic and technical development in atomic energy in many countries, intended mainly for business and industry.
3. *Atomic Energy Clearing House.* Congressional Information Bureau Inc., Mills Building, Washington 6, D.C. 1955 and onwards. (Weekly.)
   A non-technical news letter designed for business men which gives details of licences, access agreements and permits issued by the AEC to industry and includes congressional news and details of new developments in the U.S.A.
4. *Atomic Energy Law Journal.* Warren Publications Inc., 89 Beach St., Boston 11, Mass. 1959 and onwards. (Quarterly.)
   Reports current legal decisions affecting atomic energy in U.S.A., new legislation and regulations, granting of licences for reactors or waste disposal facilities and book reviews.
5. *Atomics.* 308 East James St., Barrington, Illinois. 1958 and onwards. (Bimonthly.)
   Short technical articles and news items about developments in atomic energy, particularly in America, with emphasis on the technological aspects. Mainly for engineers and management.
6. *Atomkernenergie.* Verlag Karl Thiemig KG, Pilgerheimer Straße 38, Munich 9, Germany. 1956 and onwards. (Monthly.)
   Covers all aspects of atomic energy with emphasis on reactors and nuclear engineering and contains long review articles with references. Text is in German with English abstracts.
7. *Atompraxis.* Verlag G.Braun, Karlsruhe, Germany. 1955 and onwards. (Monthly.)
   An international journal for applied nuclear science in industry, agriculture, science and medicine. Articles in English, French or German with abstracts in all three.
8. *Atomwirtschaft.* Verlag Handelsblatt G.m.b.H., Kreuzstraße 21, Düsseldorf, Germany. 1955 and onwards. (Monthly.)
   Covers all aspects of nuclear power and the industrial and medical applications of atomic energy. Technical review articles and news items in German with summaries of the longer articles in English and French. Special numbers are issued devoted to summarizing nuclear developments in particular countries.
9. *Bulletin of the Atomic Scientists.* Educations Foundation for Nuclear Science, 935 E. 60th St., Chicago 37, Illinois. 1945 and onwards. (Monthly.)

Devoted to commentary on the political, moral and social aspects of science in general and atomic energy in particular, and the relations between scientists and the community.
10. *Canadian Nuclear Technology.* Maclean-Hunter Pub. Co. Ltd., 481 University Avenue, Toronto 2, Ontario. 1961 and onwards. (Quarterly.)
The principal Canadian nuclear journal, published to inform those outside Canada of Canadian developments and to provide Canadian industry with a summary of world developments in the nuclear field.
11. *Energia és Atomtechnika.* Hungarian Technical Publishers, Bajscy Zsilinszky-ut 22, Budapest 5, Hungary. 1948 and onwards. (Monthly.)
Theoretical and technical articles on various aspects of atomic energy, particularly nuclear power. Articles are in Hungarian with titles in English, German and Russian.
12. *Energia Nuclear.* Junta de Energia Nuclear, Ciudad Universitaria 22, Madrid 3, Spain. 1957 and onwards. (Quarterly.)
Original articles and notes on all aspects of atomic energy. Covers developments in Spain and other countries. In Spanish.
13. *Energia Nucleare.* Segrate, via Redecesio 12, Milan, Italy. 1951 and onwards. (Monthly.)
Original articles and letters covering most aspects of atomic energy. Articles in Italian or English with summaries in the other language.
14. *Énergie Nucléaire.* 28 rue Saint Dominique, Paris 7e. 1959 and onwards. (8 issues per year.)
Covers nuclear physics, nuclear chemistry, nuclear engineering and general aspects of atomic energy. Articles in French with English and German abstracts.
15. *Euronuclear.* Morgan Bros., 28 Essex St., London, W.C.2. September 1964 and onwards. (Monthly.)
Technical articles on all aspects of atomic energy and particulary those which relate to the exchange of ideas and information between European countries. Text in English. Contents (with abstracts) in English, French and German. Special feature is the " News Review " which covers all countries.
16. *Europa Nucleare.* Euratom Commission of the International Business Executives Confederation, via Aniene 14, Rome, Italy. 1958 and onwards. (Bimonthly.)
Articles on technical, political, economic and international aspects of atomic energy in English, French, German or Italian with long summaries in the other three languages.
17. *Industries Atomiques.* Editions René Kister, 33 quai Wilson, Geneva. 1956 and onwards. (Bimonthly.)
International review covering all aspects of atomic energy. Articles in French with abstracts in English and German.
18. *Institute of Electrical and Electronic Engineers. Professional Technical Group on Nuclear Science. Transactions.* IEEE, 1 East 79 St., New York 21, N.Y. 1954 and onwards. (Irregular.)
Original articles on those aspects of nuclear science with which electrical or electronic engineers are concerned, particularly instrumentation and control of reactors, radiation detectors, particle accelerators, plasma physics and associated electronic instrumentation.
19. *Journal of Nuclear Science and Technology.* Atomic Energy Society of Japan, 1—1, Shiba-Tamura-cho, Tokyo, Japan. April 1964 and onwards. (Monthly.)
Original papers and short notes on Japanese research (in English) together with summaries (in English) of reports published (in Japanese) in the *Journal of the Atomic Energy Society of Japan.*
20. *Journal of Scientific Instruments.* Institute of Physics and the Physical Society, 47 Belgrave Sq., London, S.W.1. 1923 and onwards. (Monthly.)

Includes original articles and notes on radiation measuring instruments, reactor instrumentation and other instruments and on instrumentation and techniques developed during the course of research in the field of nuclear energy.

21. *Kernenergie.* VEB Deutscher Verlag der Wissenschaften, Niederwallstraße 39, 108 Berlin. 1958 and onwards. (Monthly.)
    Original papers, notes of meetings, news items and patent information relating to nuclear research in the German Democratic Republic.

22. *New Scientist.* Cromwell House, Fulwood Place, High Holborn, London W.C.1. 1956 and onwards. (Weekly.)
    Intended for non-specialists and the general public, it frequently carries review articles and short reports on the most significant developments in nuclear science and technology.

23. *Nuclear Energy.* Princes Press, 147 Victoria St., London, S.W.1. December 1959 and onwards. (Monthly.)
    A continuation of *Combustion, Boiler House and Nuclear Review* retaining the volume numbering of its predecessor. From April 1958 to November 1959 issued under the title *Nuclear Energy Engineer.* Now the official journal of the Institution of Nuclear Engineers. Devoted mainly to reactor technology, nuclear engineering and instrumentation.

24. *Nuclear Instruments and Methods.* North Holland Publishing Co., P.O. Box 103, Amsterdam. 1957 and onwards. (Monthly.)
    Original articles on accelerators, nuclear instrumentation and electronic techniques. Articles mainly in English, occasionally in French and German with English abstracts.

25. *Nuclear News.* American Nuclear Society, 244 E. Ogden Ave., Hinsdale, Illinois. 1958 and onwards. (Monthly.)
    Special articles and recent news of atomic energy developments, particularly in America.

26. *Nucleonics.* McGraw-Hill, 330 West 42nd St., New York 36, N.Y. 1947 and onwards. (Monthly.)
    Short articles and surveys on reactor physics and engineering, nuclear instrumentation and techniques, nuclear materials and irradiation.

27. *Nukleonik.* Springer Verlag, Heidelberger Platz 3, Berlin 31 (Wilmersdorf). 1958 and onwards. (Irregular.)
    Original papers and short communications on reactor technology, nuclear chemistry, radioisotopes, radiochemistry, waste disposal, radiation protection and nuclear applications in industry and biology.

28. *Review of Scientific Instruments.* American Institute of Physics, 335 East 45th St., New York 17, N.Y. 1930 and onwards. (Monthly.)
    Includes original and review articles and notes on all instrumentation aspects of atomic energy.

29. *Vacuum.* Pergamon Press, Headington Hill Hall, Oxford. 1951 and onwards. (Monthly.)
    An international journal on vacuum science and technology. Original articles and news items on developments in vacuum science including those of interest in the atomic energy field. Each issue includes a section devoted to abstracts of current literature to which annual author and subject indexes are provided.

30. *Le Vide.* Société Français des Ingenieurs et Techniciens du Vide, 147 ter A, boulevard de Strasbourg, Nogent-sur-Marne (Seine). 1946 and onwards. (Bi-monthly.)
    Original articles (in French) on vacuum techniques and applications. Each issue contains bibliography of recent literature, some items with annotations.

CHAPTER 8

ATOMIC, NUCLEAR AND HIGH ENERGY PHYSICS

ATOMIC PHYSICS

Modern physics rests largely on discoveries made during an intensely creative decade in physics ending in 1905. During this period Becquerel discovered radioactivity, J. J. Thomson demonstrated the existence of the electron, Roentgen discovered X-rays, Planck put forward the hypothesis which led to the development of quantum theory and Einstein enunciated the special theory of relativity. Together these discoveries constituted a revolution in physics and marked a dividing line between classical physics, which was largely concerned with phenomena on the macroscopic scale, and modern physics in which the fundamental structure of the universe is studied through the medium of the very small.

The earlier books on atomic physics cover all aspects of the properties and structure of atoms and do not differentiate specifically between atomic and nuclear physics. Examples are Max Born's *Atomic Physics*, [1] which was first published in Germany in 1933 under the title *Modern Physics* and is still one of the best introductions to the subject, and the *Outline of Atomic Physics*, [2] by several authors which also appeared first in 1933, based on lectures given to college students who had completed a year's course in physics prior to taking other science subjects. Even in more recent textbooks such as Shankland's *Atomic and Nuclear Physics* [3] the intimate relationship between atomic and nuclear phenomena is emphasized. Generally however the modern tendency is to separate the two subjects in the literature. *Atomic Physics* [4] by Harnwell and Stephens excludes any specific discussion of the nucleus, presenting only the basic ideas of modern atomic theory at a level suitable for physics graduates. Similarly Eisberg's *Fundamentals of Modern Physics*, [5] designed for undergraduates in their final year, discusses quantum theory and relativity theory in some detail and applies this to a more mature explanation of atomic phenomena than is usual at this level, but says little about nuclear phenomena.

For the student entering the field of nuclear science there is a number of introductory texts which cover both atomic and nuclear physics while maintaining a distinction between them. Tolansky's *Introduction to Atomic Physics* [6] covers the subject comprehensively from the physics of the electron to the principles of nuclear reactors and is based on lectures to honours students at Manchester University. A similar introduction by Semat, [7] while not quite so comprehensive, emphasizes the experimental foundations of the subject, and is designed for science and engineering graduates. Yarwood's *Atomic Physics* [8] also stresses the experimental aspects of the subject and its historical origins.

Atomic and Nuclear Physics, [9] by Littlefield and Thornley, is based on lectures given to undergraduates who may later specialize in science subjects other than physics. The treatment is largely non-mathematical so that the book can be used by students in their last year at school. A book with a similar title by Bush [10] provides a background in the theory of the subject for engineers entering the nuclear reactor field and emphasizes those aspects which have most bearing on the study of nuclear engineering. One of the best historical approaches to the subject, at roughly the same level, is *Modern Physics,* [11] by Van Name, which provides a broad survey for those with some knowledge of classical physics and elementary mathematics.

Atomic physics is not the easiest subject to describe in language suitable for the layman, although this has been attempted in a general way in a number of the books mentioned in Chapter 7 (see the section on "Books for the non-scientist"). Mention should however be made of *Atomic Physics Today* [12] by Otto Frisch, a man who played a vital role in the discovery of nuclear fission. The book is based on articles written over a period of years and conveys, to the general reader, something of the wonder and excitement that the physicist has experienced in the last half-centruy.

NUCLEAR PHYSICS
General

Between Rutherford's disintegration of the nitrogen nucleus in 1919 and the discovery of fission in 1939 a whole series of important discoveries were made, each a link in the chain leading to the release of nuclear energy. The original papers in which some of these ideas were reported have been reprinted and published in one volume [13] together with a 118-page bibliography of journal articles published up to 1947, covering all aspects of nuclear physics. From 1947 onwards the literature of the subject is well covered in *Nuclear Science Abstracts* and *Physics Abstracts* and in a number of major textbooks and monographs, some of which are described below.

An essential reference book for the nuclear physicist is *Experimental Nuclear Physics,* [14] edited by Segré, which is a review by a number of contributors of experimental work up to 1953. The three volumes are a miniature "Handbuch der Physik" with a long list of references to source material at the end of each chapter. The experimental aspects of the subject are also emphasized in *The Atomic Nucleus,* [15] by Evans, based on lectures given at M.I.T. to students who have already completed an introductory course in atomic physics, and in Kaplan's *Nuclear Physics* [16] which has been used as a standard text for introductory courses suitable for engineers and physicists with a background in classical physics and a working knowledge of calculus.

Theoretical aspects of the subject are discussed in the volume by Blatt and Weisskopf [17] which is still a standard work. This is an advanced text on low energy nuclear physics designed to give the experimental physicist a clear understanding of the theoretical background. Elton's *Introductory Nuclear Theory* [18] is also concerned with the theoretical aspects and is based on lectures given to final honours students at King's College, London. For graduates who already have a good grounding in quantum mechanics, a useful introduction is Green's *Nuclear Physics* [19] which confines itself to theories and experiments having a direct bearing on the detailed properties of the nucleus.

The physics of the nucleus is fundamental to all branches of applied nuclear science and concerns not only the nuclear physicist but every scientist, engineer and technician associated with atomic energy. One of the features of the literature in this field is the number of publications which have been written for those who are not and may never become nuclear physicists. Typical is *Basic Nuclear Physics* [20] which owes much to lectures given at the UKAEA Isotope School to students with little more than an elementary knowledge of physics. At a slightly more advanced level are Burcham's *Nuclear Physics* [21] and Halliday's *Introductory Nuclear Physics* [22] both designed for undergraduate students and both emphasizing the experimental aspects of the subject. Both theory and experiment are covered in Harvey's *Introduction to Nuclear Physics and Chemistry* [23] which is suitable for students in any scientific field and presents an outline of modern theories of nuclear structure without the use of advanced mathematics. A concise summary of the essential principles of nuclear physics is given in Mansfield's *Elementary Nuclear Physics,* [24] prepared specially for engineers who require more than a nodding acquaintance with the subject but have neither the time nor the inclination to tackle the more voluminous texts.

Nuclear Properties and Structure

A good deal of information about the structure of the nucleus has been derived from a study of the fundamental physical properties, i.e. size, mass and angular momentum. Elton's *Nuclear Sizes* [25] is a well-documented survey of experimental evidence of the size of nuclei, obtained mainly from scattering experiments. Two important conferences [26, 27] have been held on nuclear masses and their determination, the principal methods being mass spectrometry and a study of nuclear reactions, and there are a number of texts on mass spectrometers [28, 29, 30, 31] which describe their applications, not only in the measurement of nuclear masses, but in isotope separation and chemical analysis.

The subject of nuclear magnetism is discussed by Ramsey [32] and more recently, and at greater length, by Abragam [33] and Kopfermann [34] and a number of hypotheses have been suggested to explain the observed values of nuclear moments. These are discussed by Blin-Stoyle [35] in a fully documented account in which the predictions of the various theories are compared with experiment.

From a study of these properties and from information derived from nuclear reactions, a good deal of knowledge is now available on nuclear structure. This is reviewed by Preston [36] with many references to the periodical literature. There have been two international conferences [37, 38] on nuclear structure at which a vast amount of experimental work was discussed but without any significant advance being made towards a unified theory. The chief difficulty is that the nature of the forces which hold the nucleus together is not fully understood. A theoretical analysis of the available data was made by Rosenfeld [39] in 1948 and again by Sachs [40] in 1953. The problems associated with the nuclear forces are examined in an introductory text [41] by Bethe and Morrison and have been the subject of an international conference [42] held in London in 1959.

From the data available it has been possible to devise a number of theoretical models, each of which can account for some but not all, of the experimental facts. These are described and compared in *Nuclear Structure* [43] by Eisenbud and Wigner, a mainly non-mathematical discussion which has also been published as a chapter in Condon's *Handbook of Physics*. The "shell" model of the nucleus was first suggested by Meyer in 1950 and elaborated in his *Elementary Theory of Nuclear Shell Structure.* [44] A review of nuclear shell theory was made by Feenberg [45] in 1955 and a comprehensive textbook on the shell model has now appeared. [46]

More recently other models have been suggested to account for certain observed phenomena and these are discussed by Nemirovskii in *Sovremennye Modeli Atomnogo Yadra*, published in Moscow in 1960, of which no less than two translations have appeared in English. [47, 48] One of these models, the optical model, which has been useful in explaining neutron scattering at high energies, is discussed by Hodgson [49] from the point of view of the experimentalist who wants to make use of the model in his own research. The theoretical foundations for the optical model are set out in an introduction to the subject by Jones, [50] which can be regarded as a complementary work to that of Hodgson.

RADIOACTIVITY

One of the earliest comprehensive accounts of radioactivity is the classic work [51] by Rutherford, Chadwick and Ellis, in which the first proof of the existence of the nuclear atom appeared. This is still a useful source book on the general properties of radiations from naturally radioactive substances. An even earlier book by Chadwick, [52] which first appeared in 1921, has been revised by Rotblat and is a useful introduction to the subject.

For many years the standard text has been the *Manual of Radioactivity* [53] which deals comprehensively with the physical and chemical aspects of radioactivity and includes references to much of the earlier literature of the subject. More recently attention has been focussed on collecting data on radioactive decay and this is systematically presented in *Decay Schemes of Radioactive Nuclei.* [54] Decay schemes, based on data available up to 1957, are given for all radionuclides and each scheme is accompanied by a full list of references to original papers. Data on radioactive decay is also contained in many of the publications mentioned under the heading " Nuclear Data " below.

An important aspect of the work with radionuclides is the need for precise measurements of the absolute values of the activity of radioactive sources and this was the subject of an IAEA conference in 1959. The proceedings [55] form a general survey of routine methods of standardization and of new developments in absolute measuring techniques for radioactive standards.

Nuclear Reactions

The study of reactions between nuclei, and between nuclei and other particles, has provided much basic information about the properties of nuclei and is fundamental to the physics of nuclear reactors. A comprehensive review of the reaction mechanismen is given by Endt [56] and his co-authors in two volumes which include experimental surveys, theoretical papers on the dynamics of reactions and theoretical discussions of nuclear models. Radioactive decay and very high energy reactions are not covered. The dynamic aspects of nuclear reactions are discussed at greater length in *Kinematics of Nuclear Reactions,* [57] first published in Moscow in 1959.

Theoretical models suggested to account for the phenomena observed in nuclear reactions include the optical model, which has already been mentioned, [49, 50] the compound nucleus model, which applies mainly to low energy reactions, and the direct reaction model, which is important at medium and high energies. The theory of direct reactions is discussed by Tobocman [58] and by Butler and Hittmair [60] and has been the subject of an international conference. [59]

An important type of reaction, mentioned already in connection with the optical model of the nucleus, is nuclear scattering. *Quantum Theory of Scattering,* [61] by Wu and Ohmura, discusses the application of scattering theory to atomic and nuclear collisions, with many references to original papers. *Nuclear Scattering* [62] describes the physics of the phenomena, including experimental techniques, and discusses the significant results in terms of nuclear forces and structure.

From the point of view of the reactor physicist the most important nuclear reaction is, of course, fission. Early work on the fission reaction is reviewed in *Nuclear Fission and Atomic Energy* [63] which is really a general introduction to nuclear science. A more up-to-date survey is Bradley's translation [64] of the proceedings of a Russian conference held in 1956 and published originally as a supplement to *Atomnaya Energiya*. The most comprehensive review of published information on fission phenomena appeared in two University of California Radiation Laboratory reports [65] which were given only a limited circulation at the time but are to be published as part of an authoritative work on the nuclear properties of the heavy elements. [66] The foundations, assumptions and weaknesses of the various "models" of nuclear fission are discussed in some detail in a recent publication by Wilets (67) which is fully documented.

FUNDAMENTAL PARTICLES AND HIGH ENERGY PHYSICS

General

A concise introduction to the subject, for the non-specialist and graduate student, is provided by Marshak and Sudershan, [68] in a review which includes selected references to the most useful information published up to 1959. A broad picture of the field is given by Williams [69] in a book intended for research scientists engaged in elementary particle physics and a more detailed account of the theoretical methods used to describe the properties of elementary particles and explain their interactions, is given by Hamilton [70] in an advanced text which emphasizes the more firmly established theories.

Individual Particles

As a climax to experimental work extending over a number of years, the electron was eventually isolated by J. J. Thomson in 1897 and some of its fundamental properties determined. The story of the events leading up to the discovery is told in *The Discovery of the Electron,* [71] a case history in scientific method, which discusses the conceptual developments that led to an acceptance of the "atomic" nature of electricity. Among the earliest works on the properties of electrons are Thomson's *Conduction of Electricity through Gases* [72] and Millikan's *The Electron,* [73] both of which are accounts of developments in which the authors played a major part. A later book [74] by Millikan, in which he describes the first measurement of the electron charge, is in part a revision of the earlier work, with additional matter on other elementary particles.

Photons and Electrons [75] introduces the student of physics to the general theory of the interaction of electromagnetic radiation with electrons and *Electron Physics* [76] is a comprehensive study of the physics of the free electron. The first part, on electron optics, is mainly of interest to electronic engineers; the second part covers the physical properties of the electron.

In reactor engineering the most interesting particle is the neutron which is vital to the maintenance of the fission chain reaction. *Introduction to Neutron Physics* [77] reviews all aspects of experimental neutron physics including the interaction between neutrons and matter. This aspect, particularly the phenomena associated with the migration of neutrons in bulk matter, is of prime importance in nuclear reactor theory and is discussed in detail in *Neutron Transport Theory,* [78] an advanced text concerned mainly with the mathematical methods used. A more concise introduction to the mathematical treatment will be found in a second book with the same title by Tait. [79]

Neutrons, being uncharged, can penetrate matter fairly easily and can be diffracted and scattered in a manner analogous to light passing through a transparent medium. They are, therefore, a useful means of investigating the structure of bulk matter and have been used, for example, to study the molecular structure of crystalline solids. The optical properties of neutrons, particularly scattering, are discussed by Hughes [80] and by Bacon, [81] both of whom deal with fundamental principles and applications.

In 1930 only three fundamental particles were known; the photon, electron and proton. To these were later added the neutron, the neutrino and two "anti-particles", the positron and the anti-neutrino. With the exception of the neutron all these are stable in the free state. The neutrino has been the subject of a good deal of experimental research and theoretical discussion, which is reviewed in Allen's *The Neutrino,* [82] an attempt to interpret significant results up to 1957 in terms of existing theories. The advent of the positron generated a completely new concept in physics, that of anti-matter, as well as introducing a hypothesis which had never before been seriously entertained, namely the creation of matter from energy, and if these ideas were accepted with reluctance by many physicists, they were even more disturbing from a philosophical point of view. The implications of these ideas are discussed by Hanson in his *Concept of the Positron* [83] which is basically an analysis of the conceptual structure of modern physics.

High Energy Physics

When the stable particles acquire a large kinetic energy, either as components of the cosmic radiation or as projectiles in particle accelerators, reactions occur in which other particles and anti-particles are produced, most of them unstable but all of them fundamental in the sense that they do not appear to be composite structures of other particles. Altogether over thirty fundamental particles and anti-particles are now known and no doubt others will be discovered.

The earliest experiments in this field were carried out through the medium of cosmic rays and concentrated mainly on the study of mu-mesons and pi-mesons. Rossi's *High Energy Particles,* [84] a standard text which gives a comprehensive account of high energy phenomena, is based on experiments of this kind. Thorndike's *Mesons* [85] is a student's introduction to the subject and *Meson Physics,* [86] by Marshak, surveys experiments up to 1951 with many references to original papers published in the preceeding five years. *High Energy Nuclear Physics* [87] by Lock, is based on a series of lectures to post-graduate students and is concerned mainly with pi-mesons and the

Yukawa theory of nuclear forces. Nishijima's *Fundamental Particles*, [88] also deals with mu-mesons and pi-mesons, in the form of lecture notes which provide a survey of the more theoretical aspects of the subject.

From 1953 onwards, as the large accelerators began to come into operation, the emphasis shifted to the "strange" particles, the K-mesons and hyperons which had already been observed in cosmic ray studies. *The Physics of Elementary Particles*, [89] by Jackson, is an introductory account for nuclear physicists, which covers pi-mesons and decay processes as well as the strange particles. Adair and Fowler [90] present a review of the experimental evidence for strange particles, summarizing their important properties, and these are discussed at greater lenght by Dalitz [91] in a comprehensive survey of the observed phenomena, which refers extensively to original papers. Most of the relevant journal articles on heavy mesons and hyperons, published up to 1958, are listed in a comprehensive bibliography by Vitale [92] which covers both theoretical and experimental aspects.

There have been a number of conferences on high energy physics including the series of Rochester conferences, [93] which were held annually at Rochester University up to 1957. In that year the International Union of Pure and Applied Physics created a Commission on High Energy Physics to organize the conference as an international event and the location of the conference is now rotated between Rochester, CERN and Kiev.

Some of the concepts associated with high energy physics are by no means easy to come to grips with, yet the subject has excited a great deal of interest among non-scientists. The cost of the next generation of particle accelerators is likely to be so high that a decision to spend the money, at least in western countries, can be made only by the elected representatives of the people, so that there is a need for the man-in-the-street to know something about the scientific issues at stake. A fairly simple account is given by Hill in *Tracking Down Particles*, [94] which discusses the influence that particle physics has had on science and technology and describes the machines and the experimental techniques which have been used. Readers with a little scientific background will find an interesting presentation in Yang's *Elementary Particles*, [95] an historical account by a recent Nobel Prize winner.

Cosmic Rays

The most complete survey of cosmic ray research is Jánossy's *Cosmic Rays*, [96] a critical exposition of experiment and theory which has become a standard text. Experimental work up to 1949 is reviewed by Montgomery [97] with selected references to original papers. Much of the early information on elementary particles was obtained through cosmic ray studies and Ramakrishnan [98] provides a mathematical account of the particle interactions observed in cosmic ray phenomena. The most energetic of the cosmic ray particles generate multiple branching reactions known as cascades, or air showers, and this phenomena is discussed in some detail by Galbraith. [99]

Not all cosmic ray research is concerned with elementary particles and indeed, now that high speed particles can be provided more easily by particle accelerators, interest is reverting to the astronomical aspects. *Cosmic Rays,* [100] by Cranshaw, is concerned only with the origin and nature of the radiation, and a book with the same title by Wolfendale, [101] although surveying the whole field at undergraduate level, emphasizes the astronomical side of cosmic ray research.

There are a number of introductions to the subject, some suitable for the general reader. Leprince-Ringuet [102] gives a fairly elementary survey up to 1950, with emphasis on the experimental techniques, and an introduction for students and non-specialists is provided by *The Cosmic Radiation,* [103] by Hooper and Scharff, which gives adequate guidance for further reading.

Particle Accelerators

Many of the general textbooks on atomic and nuclear physics devote a section to particle accelerators and the most useful of these are starred in the list of references at the end of this chapter. Ten years ago very little was available on this subject outside of journal articles and conference papers but, fortunately, this is no longer the case.

One of the earliest introductions to the techniques of particle acceleration was a publication [104] based on four papers given at an Institute of Physics Convention in 1949, which described developments up to that time, with a number of references to the earlier literature. This had been preceeded in 1945 by Mann's monograph [105] on the cyclotron which was originally an account of the design and operation of the first large-scale cyclotron, the 37-inch machine at Berkeley; in later editions an additional chapter was added to round off the story and to provide references to papers on more recent machines which the author mentions only in passing.

The first general account of high energy machines is contained in Livingston's *High Energy Accelerators* [106] which is an introduction to the design of machines producing particles with energies greater than about 200 MeV and their application in nuclear and elementary particle physics. This was followed in 1962 by *Particle Accelerators,* [107] which is a critical and comprehensive analysis of the physical principles of all types of accelerators and their associated equipment; a fully documented text for engineers and physicists which has become a standard work.

For the student with little more mathematical equipment than basic calculus, one of the best introductions to the subject is Livingood's *Principles of Cyclic Particle Accelerators* [108] which covers every kind of machine except those of the Cockcroft–Walton and Van de Graaff type and includes a long annotated bibliography of original papers and books. The basic problems involved in the theory and design of linear ion accelerators are discussed in a recent Russian publication, [109] now available in translation; earlier papers on linear accelerators are listed, with abstracts, in a USAEC bibliography [110] issued in 1958.

One of the earliest books on accelerators written for the layman was Solomon's *Why Smash Atoms?* [111] which, even in the revised edition, does not include modern developments but is an easily understood introduction to basic principles. A more recent book [112] by Wilson and Littauer is designed for secondary school students and laymen with some scientific knowledge, and includes developments up to 1960. Ratner's *Accelerators of Charged Particles* [113] gives a Russian viewpoint on the design and purpose of accelerators and is well within the layman's understanding.

For the non-specialist a bibliography [114] of the principal books and review articles issued up to July 1962 has been compiled by the Atomic Energy Research Establishment, Harwell and issued by the Information Office as a reading list.

NUCLEAR DATA

A significant proportion of atomic and nuclear research is concerned with the acquisition and analysis of data relating to atomic and nuclear properties and reactions. Experimental results reported in the literature frequently show significant discrepancies and a search for "best values" may in some cases take longer than an actual experiment. To aid the physicist a number of tabulations of nuclear data have been undertaken, some of them on a continuing basis, and the most useful of these are listed in the *Directory to Nuclear Data Tabulations* [115] issued by the Nuclear Data Project of the U.S. National Academy of Sciences. The *Directory* includes compilations of experimental and theoretical data in basic nuclear physics and chemistry but is not critical, in the sense that no attempt is made to characterize any table as containing best values of a particular property.

Nuclear data has been published by the USAEC since 1952 in *New Nuclear Data*, [116] issued as a supplement to *Nuclear Science Abstracts*. In 1959 this publication was superseded by *Nuclear Data Tables*, [117, 118] each issue of which contains a supplement to the *Directory* mentioned above, as well as new tables of properties.

One of the most comprehensive collections of data is *Tabellen der Atomkerne,* first published in Berlin in 1958 and now available in translation. [119] The tables contain a vast amount of information on nuclear properties and are an essential tool for the nuclear physicist. Another important tabulation is the *Table of Isotopes,* [120] a complete list of all the radioactive and stable isotopes of the elements and their properties as recorded in the literature up to February 1958. The data in this table relating to gamma-emitting nuclides has been extracted and rearranged in a tabulation by Slater [121] for the use of those concerned with the identification and elucidation of gamma-ray spectra.

Basic to atomic and nuclear physics are the atomic constants; the mass and charge of the electron, the velocity of light, Avogadro's number, the mass of the proton, the proton magnetic moment and Planck's constant. *Fundamental Atomic Constants* [122] gives a concise and fully documented account of the constants, their earlier and recent determination and discusses the considerations involved in the choice of best values.

Apart from the Nuclear Data Project of the National Academy of Sciences there are a number of centres specializing in the collection and analysis of nuclear data, some of which have already been mentioned in earlier chapters. A list of the more important of these is given in an article [123] in *Physics Today* which also discusses the work of some of these centres.

REFERENCES

Atomic Physics

1. BORN, M. *Atomic physics.* 7th ed. London, Blackie, 1962.
2. BLACKWOOD, O. H. AND OTHERS. *An outline of atomic physics.* 3rd ed. New York, Wiley; London, Chapman and Hall, 1955.
3. SHANKLAND, R. S. *Atomic and nuclear physics.* New York, London, Macmillan, 1955.
4. HARNWELL, G. P. AND STEPHENS, W. E. *Atomic physics.* New York, McGraw-Hill, 1955.
5. EISBERG, R. M. *Fundamentals of modern physics.* New York, Wiley. 1961.
6. *TOLANSKY, S. *Introduction to atomic physics.* 5th ed. London, Longmans, 1963.
7. *SEMAT, H. *Introduction to atomic and nuclear physics.* 4th ed. London, Chapman and Hall, 1962.
8. *YARWOOD, J. *Atomic physics.* 2nd ed. London, University Tutorial Press. 1963. (Vol. 2 of "Electricity, Magnetism and Atomic Physics".)
9. *LITTLEFIELD, T. A. AND THORNLEY, N. *Atomic and nuclear physics.* London, Van Nostrand, 1963.
10. BUSH, H. D. *Atomic and nuclear physics: theoretical principles.* Englewood Cliffs (N.J.), Prentice-Hall; London, Iliffe, 1962.
11. VAN NAME, F. W. *Modern physics.* 2nd ed. Englewood Cliffs (N.J.), London, Prentice-Hall, 1962.
12. FRISCH, O. R. *Atomic physics today.* New York, Basic Books Inc., 1961. London, Oliver and Boyd, 1962.

Nuclear Physics

13. BEYER, R. T., ED. *Foundations of nuclear physics.* New York, Dover Publications, 1949.
14. *SEGRÉ, E., ED. *Experimental nuclear physics.* New York, Wiley; London, Chapman and Hall. 3 vols. Vols. 1 and 2, 1953. Vol. 3, 1959.
15. EVANS, R. D. *The atomic nucleus.* New York, McGraw-Hill, 1955.
16. *KAPLAN, I. *Nuclear physics.* 2nd ed. Reading (Mass.), London, Addison-Wesley, 1963.
17. BLATT, J. M. AND WEISSKOPF, V. F. *Theoretical nuclear physics.* New York, Wiley; London, Chapman and Hall, 1952.
18. ELTON, L. R. B. *Introductory nuclear theory.* London, Pitman, 1959.
19. *GREEN, A. E. S. *Nuclear physics.* New York, McGraw-Hill, 1955.

20. *WILLIAMS, I. R. AND WILLIAMS, M. W. *Basic nuclear physics.* London, Newnes, 1962.
21. *BURCHAM, W. E. *Nuclear physics: an introduction.* London, Longmans, 1963.
22. *HALLIDAY, D. *Introductory nuclear physics.* 2nd ed. New York, Wiley; London, Chapman and Hall, 1955.
23. *HARVEY, B. G. *Introduction to nuclear physics and chemistry.* Englewood Cliffs (N.J.), London, Prentice-Hall, 1962.
24. *MANSFIELD, W. K. *Elementary nuclear physics.* London, Temple Press, 1958. (Nuclear Engineering Monographs.)

Nuclear Properties and Structure

25. ELTON, L. R. B. *Nuclear sizes.* London, O.U.P., 1961.
26. HINTENBERGER, H., ED. *Nuclear masses and their determination: proceedings of the Conference on the Precision Determination of Atomic Nuclear Masses, Mainz, July, 1956.* London, Pergamon Press, 1957.
27. DUCKWORTH, H. E., ED. *Proceedings of the International Conference on Nuclidic Masses, McMaster University, Hamilton, Ontario, September, 1960.* Toronto, University of Toronto Press, 1960.
28. BARNARD, G. P. *Modern mass spectrometry.* London, Institute of Physics, 1953.
29. DUCKWORTH, H. E. *Mass spectroscopy.* London, C.U.P., 1958.
30. ROBERTSON, H. J. *Mass spectrometry.* London, Methuen, 1959.
31. McDOWELL, C. A. *Mass spectrometry.* New York, London, McGraw-Hill, 1963.
32. RAMSEY, N. F. *Nuclear magnetic moments.* New York, Wiley, 1953.
33. ABRAGAM, A. *Principles of nuclear magnetism.* Oxford, Clarendon Press, 1961.
34. KOPFERMANN, H. *Nuclear moments,* English version prepared from the 2nd German edition by E. E. Schneider. New York, Academic Press, 1958.
35. BLIN-STOYLE, R. G. *Theories of nuclear moments.* London, O.U.P., 1957.
36. PRESTON, M. A. *Physics of the nucleus.* Reading (Mass.), London, Addison-Wesley, 1962.
37. LIPKIN, H. J., ED. *Proceedings of the Rehovoth Conference on Nuclear Structure held at the Weizmann Institute of Science, Rehovoth, September, 1957.* Amsterdam, North Holland Pub. Co., 1958.
38. BROMLEY, D. A. AND VOGT, E. W., EDS. *Proceedings of the International Conference on Nuclear Structure, Kingston, Ontario, August/September, 1960.* Toronto, University of Toronto Press; Amsterdam, North Holland Pub. Co., 1960.
39. ROSENFELD, L. *Nuclear forces.* Amsterdam, North Holland Pub. Co., 1948.
40. SACHS, R. G. *Nuclear theory.* Cambridge (Mass.), Addison-Wesley, 1953.
41. BETHE, H. A. AND MORRISON, P. *Elementary nuclear theory.* 2nd ed. New York, Wiley, 1956.
42. GRIFFITH, T. C. AND POWER, E. A., EDS. *Nuclear forces and the few-nucleon problem: proceedings of the international conference held at University College, London, July, 1959.* Oxford, Pergamon Press, 1960. 2 vols.
43. EISENBUD, L. AND WIGNER, E. P. *Nuclear structure.* London, O.U.P., 1958.
44. MEYER, M. G. AND JENSEN, J. H. D. *Elementary theory of nuclear shell structure.* New York, Wiley; London, Chapman and Hall, 1955.
45. FEENBERG, E. *Shell theory of the nucleus.* Princeton University Press; London, O.U.P., 1955.
46. DE-SHALIT, A. AND TALMI, I. *Nuclear shell theory.* New York, London, Academic Press, 1963.
47. NEMIROVSKII, P. E. *Contemporary models of the atomic nucleus,* translated from the Russian by S. and M. Nikolić. Oxford, Pergamon Press, 1963.
48. NEMIROVSKII, P. E. *Nuclear models,* translated by S. Chomet. London, E. and F. Spon, 1963.

49. HODGSON, P. E. *The optical model of elastic scattering.* Oxford, Clarendon Press, 1963.
50. JONES, P. B. *Optical model in nuclear and particle physics.* New York, London, Interscience, 1963.

Radioactivity

51. RUTHERFORD, E. J., CHADWICK, J. AND ELLIS, C. D. *Radiations from radioactive substances.* London, C.U.P., 1930.
52. CHADWICK, SIR J. *Radioactivity and radioactive substances.* 4th ed. rev. and supplemented by Professor J. Rotblat. London, Pitman, 1961. (4th edition originally published 1953.)
53. HEVESY, G. AND PANETH, F. A. *A manual of radioactivity.* 2nd ed. enlarged and revised. London, O.U.P., 1938.
54. DZHELEPOV, B. S. AND PEKER, L. K. *Decay schemes of radioactive nuclei.* Oxford, Pergamon Press, 1961.
55. INTERNATIONAL ATOMIC ENERGY AGENCY. *Metrology of radionuclides: proceedings of a symposium organised by the IAEA, Vienna, October, 1959.* Vienna, IAEA, 1960.

Nuclear Reactions

56. ENDT, P. M. AND OTHERS. *Nuclear reactions.* Amsterdam, North Holland Pub. Co. 2 vols.
Vol. 1, by Endt, P. M. and Demeur, M., 1959; Vol. 2, by Endt, P. M. and Smith, P. B., 1962.
57. BALDIN, A. M. AND OTHERS. *Kinematics of nuclear reactions,* translated from the Russian by W. E. Jones. Oxford, Pergamon Press, 1961. (First published in Moscow in 1959.)
58. TOBOCMAN, W. *Theory of direct nuclear reactions.* London, O.U.P., 1961.
59. CLEMENTEL, E. AND VILLI, C., EDS. *Proceedings of the Conference on Direct Reactions and Nuclear Reaction Mechanisms, held at the University of Padua, September, 1962.* New York, London, Gordon and Breach, 1963.
60. BUTLER, S. T. AND HITTMAIR, O. H. *Nuclear stripping reactions.* New York, Wiley; London, Pitman, 1957.
61. WU, TA-YOU AND OHMURA, T. *Quantum theory of scattering.* London, Prentice-Hall, 1962.
62. MATHER, K. B. AND SWAN, P. *Nuclear scattering.* London, C.U.P., 1958.
63. STEPHENS, W. E., ED. *Nuclear fission and atomic energy.* Lancaster (Pa.), The Science Press, 1948.
64. BRADLEY, J. E. S.,TRANS. *Physics of nuclear fission,* translation of supplement 1 of the Soviet journal *Atomnaya Energiya.* Oxford, Pergamon Press, 1958.
65. HYDE, E. K. *A review of nuclear fission.* University of California, Lawrence Radiation Laboratory, 1960.
Part. 1: Fission phenomena at low energy. UCRL 9036, January 1960. Part. 2: Fission phenomena at moderate and high energy. UCRL 9065, February 1960.
66. HYDE, E. K., PERLMAN, I. AND SEABORG, G. T. *Nuclear properties of the heavy elements.* Englewood Cliffs (N.J.), London, Prentice-Hall, 3 vols. (To be published in January 1965.)
67. WILETS, L. *Theories of nuclear fission.* Oxford, Clarendon Press, 1964.

Elementary Particles and High Energy Physics

68. MARSHAK, R. E. AND SUDERSHAN, E. C. G. *Introduction to elementary particle physics.* New York, London, Interscience, 1961.

69. WILLIAMS, W. S. C. *An introduction to elementary particles*. New York, London, Academic Press, 1961.
70. HAMILTON, J. *The theory of elementary particles*. Oxford, Clarendon Press, 1959.
71. ANDERSON, D. L. *The discovery of the electron: the development of the atomic concept of electricity*. Princeton, Van Nostrand, 1964.
72. THOMSON, J. J. *The conduction of electricity through gases*. 3rd ed. London, C.U.P., 1928. (2nd ed. published 1906.)
73. MILLIKAN, R. A. *The electron: its isolation and measurement and the determination of some of its properties*. Chicago University Press, 1964. (Facsimile reprint of 1917 ed.)
74. MILLIKAN, R. A. *Electrons (+ and —), protons, photons, neutrons, mesotrons and cosmic rays*. Rev. ed. Chicago University Press, 1947. (1st ed. 1935.)
75. SPRING, K. H. *Photons and electrons*. 2nd ed. London, Methuen; New York, Wiley, 1960.
76. KLEMPERER, O. *Electron physics: the physics of the free electron*. London, Butterworth, 1959.
77. CURTISS, L. F. *Introduction to neutron physics*. Princeton, Van Nostrand, 1959.
78. DAVISON, B. AND SYKES, J. B. *Neutron transport theory*. Oxford, Clarendon Press, 1957.
79. TAIT, J. H. *Neutron transport theory*. London, Longmans, 1964.
80. HUGHES, D. J. *Neutron optics*. New York, London, Interscience, 1954.
81. BACON, G. E. *Neutron diffraction*. 2nd ed. Oxford, Clarendon Press, 1962.
82. ALLEN, J. S. *The neutrino*. Princeton University Press; London, O.U.P., 1958.
83. HANSON, N. R. *The concept of the positron: a philosophical analysis*. London, C.U.P., 1963.
84. ROSSI, B. *High energy particles*. New York, Prentice-Hall, 1952.
85. THORNDIKE, A. M. *Mesons: a summary of experimental facts*. New York, McGraw-Hill, 1952.
86. MARSHAK, R. E. *Meson physics*. New York, McGraw-Hill, 1952.
87. LOCK, W. O. *High energy nuclear physics*. London, Methuen; New York, Wiley, 1960.
88. NISHIJIMA, K. *Fundamental particles*. New York, W. A. Benjamin, 1963.
89. JACKSON, J. D. *The physics of elementary particles*. Princeton University Press; London, O.U.P., 1958.
90. ADAIR, R. K. AND FOWLER, E. C. *Strange particles*. New York, London, Interscience, 1963.
91. DALITZ, R. H. *Strange particles and strong interactions*. London, O.U.P., 1962.
92. VITALE, B. *A bibliography on heavy mesons and hyperons*. Amsterdam, North Holland, 1960.
93. *High energy nuclear physics; proceedings of the annual Rochester Conference.* Published by the University of Rochester and distributed by Interscience, 1952 to 1957.
The 1st and 2nd conferences were held in 1950 and 1951 respectively but no proceedings were issued. Proceedings were issued as above for the 3rd, 4th, 5th, 6th and 7th conferences held 1952 to 1957. Thereafter proceedings were issued as follows:
1958 Annual International Conference on High Energy Physics at CERN, Geneva, June—July, 1958. Proceedings, ed. by B. Ferretti. Geneva, CERN, 1958. (This is the 8th conference.)
1960 Annual International Conference on High Energy Physics at Rochester, August—September, 1960. Proceedings, ed. by E. C. G. Sudershan and others. University of Rochester and Interscience, 1960. (This is the 10th conference.)
1962 International Conference on High Energy Physics at CERN, Geneva,

July 1962. Proceedings, ed. by J. Prentki. Geneva, CERN, 1962. (This is the 11th conference.)
The 9th conference was held at Kiev in 1959 but no proceedings were issued.
94. *HILL, R. D. *Tracking down particles.* New York, W. A. Benjamin, 1963.
95. YANG, C. N. *Elementary particles: a short history of some discoveries in atomic physics.* Princeton University Press, 1961. London, O.U.P., 1962.

Cosmic Rays

96. JANOSSY, L. *Cosmic rays.* Oxford, Clarendon Press, 1948.
97. MONTGOMERY, D. J. *Cosmic ray physics.* Princeton University Press, 1949.
98. RAMAKRISHNAN, A. *Elementary particles and cosmic rays.* Oxford, Pergamon Press, 1962.
99. GALBRAITH, W. *Extensive air showers.* London, Butterworth, 1958.
100. CRANSHAW, T. E. *Cosmic rays.* Oxford, Clarendon Press, 1963.
101. WOLFENDALE, A. W. *Cosmic rays.* London, Newnes, 1963.
102. LEPRINCE-RINGUET, L. *Cosmic rays.* New York, Prentice-Hall, 1950.
103. HOOPER, J. E. AND SCHARFF, M. *The cosmic radiation.* London, Methuen; New York, Wiley, 1958.

Particle Accelerators

104. ROTBLAT, J. AND OTHERS. *The acceleration of particles to high energies: based on a session arranged by the Electronics Group at the Institute of Physics Convention, May, 1949.* London, Institute of Physics, 1950.
105. MANN, W. B. *The cyclotron.* 4th ed. London, Methuen; New York. Wiley, 1953.
106. LIVINGSTON, M. S. *High energy accelerators.* New York, Interscience, 1954.
107. LIVINGSTON, M. S. AND BLEWETT, J. P. *Particle accelerators.* New York, London, McGraw-Hill, 1962.
108. LIVINGOOD, J. J. *Principles of cyclic particle accelerators.* Princeton, Van Nostrand, 1961.
109. KARETNIKOV, D. V. AND OTHERS. *Linear ion accelerators.* Moscow, Gosatomizdat, 1962. (Translation published as AEC—tr—6229, Washington, Office of Technical Services, 1964.)
110. MALMBERG, C., COMP. *Abstract bibliography on linear accelerators.* Washington, Office of Technical Services, March 1958. (AECU 4009.)
Supplements: Bibliography on travelling wave linear accelerators, April 1958. (AECU 4010.) Bibliography on standing wave linear accelerators, June 1958. (AECU 4012.)
111. SOLOMON, A. K. *Why smash atoms?* Rev. ed. Harmondsworth, Penguin Books, 1959. (First published in 1940.)
112. WILSON, R. R. AND LITTAUER, R. *Accelerators: machines of nuclear physics.* New York, Anchor Books, 1960; London, Heinmann, 1962.
113. RATNER, B. S. *Accelerators of charged particles,* translated from the Russian by L. A. Fenn. Oxford, Pergamon Press, 1964.
114. SKEATS, N. B. AND MULLINS, P. R. *Particle accelerators: selected references to books and articles describing the principles and purposes of different types of accelerators.* Information Office, Atomic Energy Research Establishment, Harwell, July 1962. (AERE Reading List 7.)

Nuclear Data

115. GIBBS, R. C. AND WAY, K., EDS. *A directory to nuclear data tabulations.* Published by the Nuclear Data Project, National Research Council, National Academy of Sciences. Washington, U.S. Government Printing Office, 1958.

First supplement covering December 1957 to December 1958, included in 1959 Nuclear Data Tables. Second supplement covering December 1958 to June 1961 included in 1960 Nuclear Data Tables, volume 4.

116. *New Nuclear Data (supplement to Nuclear Science Abstracts)*. 1952, NSA, vol. 6, no. 24 B; 1953, NSA, vol. 7, no. 24 B; 1954, NSA, vol. 8, no. 24 B; 1955, NSA, vol. 9, no. 24 B; 1956, NSA, vol. 10, no. 24 B. New Nuclear Data: 1957 Cumulation. (Published separately by USAEC, November 1956.)
117. WAY, K., ED. *1959 Nuclear data tables*. Washington, U.S. Government Printing Office, April 1959.
118. WAY, K., ED. *1960 Nuclear data tables*. Washington, U.S. Government Printing Office, 1961. 4 vols.
119. KUNZ, W. AND SCHINTLMEISTER, J. P. *Nuclear tables. Part 1. Nuclear properties of the elements*. Oxford, Pergamon Press, 1963. 2 vols. First published as Tabellen der Atomkerne, Berlin, Akademie Verlag, 1958. A second part is to be published dealing with nuclear reaction data.
120. STROMINGER, D., HOLLANDER, J. M. AND SEABORG, G. T. Table of isotopes. *Revs. Mod. Phys.*, vol. 30, no. 2, pt. 2, pp. 585–904, April 1958. (See also reference 54.)
121. SLATER, D. N. *Gamma-ray radionuclides in order of increasing energy*. London, Butterworth, 1962.
122. SANDERS, J. H. *Fundamental atomic constants*. London, O.U.P., 1961.
123. GOVE, N. B. Information centres in nuclear physics. *Physics Today*, pp. 44–47. January 1964.

APPENDIX

Periodicals which Contain Articles on Nuclear and High Energy Physics

1. *Annals of Physics*. Academic Press, 111 Fifth Avenue, New York 3, N.Y., 1957 and onwards. (Monthly.)
   Original papers on fundamental research in physics. A high proportion are on nuclear physics and elementary particle physics.
2. *British Journal of Applied Physics*. Institute of Physics and the Physical Society, 47 Belgrave Square, London, S.W.1., 1950 and onwards. (Monthly.)
   Original papers and research notes on theoretical and experimental aspects of applied physics, including properties of materials, wave propagation, semiconductors and dielectrics, and applications of nuclear, solid state, and plasma physics.
3. *Bulletin of the American Physical Society*. American Physical Society, Columbia University, New York. Series 2, 1956 and onwards. (7 issues a year.)
   The Bulletin contains only abstracts of papers which are to be presented at meetings of the American Physical Society. In general these papers are not available at the time the abstract appears and further information can be obtained only by a direct approach to authors.
4. *Canadian Journal of Physics*. National Research Council, Sussex Drive, Ottawa 2, 1929 and onwards. (Monthly.)
   Original papers on fundamental and applied aspects of physics including nuclear physics.
5. *Comptes Rendus Hebdomadaires des Séances de l'Académie des Sciences*. Gauthier-Villars, 55 quai des Grands-Augustins, Paris 6e, 1835 and onwards. (Weekly.)
   The major French scientific journal covering all sciences and containing original papers and communications on nuclear and high energy physics.

6. *Contemporary Physics.* Taylor and Francis, Red Lion Court, London, E.C.4, 1959 and onwards. (Bimonthly.)
   Review articles reporting developments in modern physics written primarily for the non-specialist.
7. *Nuclear Physics.* North Holland Publishing Co., P.O. Box 103, Amsterdam, 1956 and onwards. (Irregular.)
   Original articles on theoretical and experimental studies of the fundamental constituents of matter and their interactions. Articles in English, French or German. Abstracts of articles in English.
8. *Nuovo Cimento.* Nicola Zanichelli, via Irnerio 34, Bologna, Italy, 1855 and onwards. (Semi-monthly.)
   Original research papers in all fields of physics, particularly nuclear and elementary particle physics. An international review issued under the auspices of the Italian Physical Society. The bulk of the articles are in English.
9. *Physical Review.* American Institute of Physics, 335 East 45th St., New York, N.Y., 1893 and onwards. (Weekly.)
   The world's most quoted journal in experimental and theoretical physics. A large proportion of the articles are concerned with nuclear and atomic physics.
10. *Physical Review Letters.* American Physical Society, Columbia University, New York, N.Y., 1958 and onwards. (Weekly.)
    Designed to allow physicists to report briefly on important results in a form which can be issued more rapidly than formal papers. Apart from the letters each issue contains short abstracts of forthcoming papers in *Physical Review.*
11. *Physics Letters.* North Holland Publishing Co., P.O. Box 103, Amsterdam, 1962 and onwards. (Semi-monthly.)
    The European equivalent of *Physical Review Letters.* Reports important results in theoretical and experimental physics including nuclear and high energy physics.
12. *Proceedings of the Physical Society.* Institute of Physics and the Physical Society, 47 Belgrave Square, London, S.W.1., 1874 and onwards. (Monthly.)
    The principal physics journal in the United Kingdom. Original papers on nuclear physics, quantum mechanics, and theoretical and experimental studies on the structure of matter.
13. *Proceedings of the Royal Society. Series A, Mathematics and the Physical Sciences.* Royal Society, Burlington House, Piccadilly, London, W.1., 1800 and onwards. (Semi-monthly.)
    Original papers containing substantial additions to scientific knowledge. The proportion of articles on nuclear physics is small; the majority are in the fields of solid state physics and theoretical physics.
14. *Progress in Elementary Particle and Cosmic Ray Physics.* North Holland Publishing Co., P.O. Box 103, Amsterdam, 1957 and onwards. (Annual.)
    Long review articles on important developments in cosmic ray research and high energy physics.
15. *Progress in Nuclear Physics.* Pergamon Press, Headington Hill Hall, Oxford, 1950 and onwards. (Annual.)
    Review articles summarizing developments in specific fields of nuclear and particle physics.
16. *Progress of Theoretical Physics.* Yukawa Hall, Kyoto University, Kyoto, Japan, 1946 and onwards. (Monthly.)
    Published by the Physical Society of Japan. Articles and letters (in English) on theoretical physics and theoretical aspects of nuclear physics.
17. *Reports on Progress in Physics.* Institute of Physics and the Physical Society, 47 Belgrave Square, London, S.W.1., 1934 and onwards. (Annual.)
    Long review articles on developments in all fields of physics.

18. *Reviews of Modern Physics.* American Institute of Physics, 335 East 45th St., New York, N.Y., 1929 and onwards. (Quarterly.)
    Comprehensive and timely discussions of problems of interest in physics and long bibliographies of source material. Frequently includes proceedings of specialized conferences in physics.
19. *Zeitschrift für Naturforschung (A), Astrophysik, Physik, Physikalische Chemie.* Verlag der Zeitschrift für Naturforschung, Postfach 61, Tübingen, 1946 and onwards. (Monthly.)
    Original articles and short communications on research in the subjects covered. Articles in German or English.
20. *Zeitschrift für Physik, Series A.* Springer-Verlag, Heidelberger Platz 3, Berlin 31 (Wilmersdorf), 1920 and onwards. (4 vols. per year.)
    Original articles on research in physics including nuclear physics.

CHAPTER 9

NUCLEAR POWER AND ENGINEERING

ENERGY RESOURCES AND NUCLEAR POWER PROGRAMMES

The energy derived from nuclear fission is but one of a number of sources available to supply the power which the world needs and the future development of atomic energy is indissolubly linked with the general problem of energy resources and requirements. There have been a number of attempts to measure the extent of this problem starting with Putnam's *Energy in the Future* [1] which contained the first serious estimate of world fuel resources and consumption rates in the foreseeable future. Atomic energy was brought into the picture in Thirring's *Power Production* [2] which is a survey of the relative abundance and accessibility of existing resources and of the demands likely to be made on them. Anyone who feels that Thirring is presenting an unnecessarily gloomy picture will not be reassured by a more recent report, [3] prepared for the USAEC, which estimates that, if the present rate of increase in consumption continues, the fossil fuel reserves will be exhausted by the year 2050, which is within the life span of children now being born. The main purpose of this report, however, was to estimate the date by which nuclear power would be needed on a large scale in the U.S.A., and in the world, if the increasing energy demands are to be met.

For the general reader, a more easily digested treatment will be found in *The Atom and the Energy Revolution,* [4] published as a Penguin Special, which begins with a review of world energy resources and requirements and goes on to deal with the nuclear power programmes of various countries. In this respect the book is a little out of date since there have been rapid developments since 1958, and some reference has already been made to these in earlier chapters.

In the United States, a country with plentiful supplies of conventional fuels, there has been no immediate need to embark on a large-scale power programme but a number of studies have been made of future requirements. In 1962 President Kennedy asked the USAEC to take "a new and hard look at the role of nuclear power in our economy", as a result of which a report was prepared which examined the economic aspects of nuclear energy production in relation to other forms of energy and outlined a tentative programme. The appendices to this report were published by the USAEC [5] and give all the facts and figures on which the report was based. A summary [6] of information on present AEC programmes for the development of nuclear power systems for electricity generation and for propulsion was issued in 1963, at the same time as a book [7] which examines in some detail the economic and political issues which have influenced U.S. policy in the development of a nuclear power programme.

In the Western world the earliest and largest nuclear power programme to be implemented is that of the United Kingdom. The first programme was announced in a white paper [8] in 1955, and called for twelve power stations producing between 1500 and 2000 megawatts of electricity by 1965. In 1957, under the spur of the Suez crisis, the programme was trebled [9] and in 1960, cut by half [10] as a result of changes, not only in the world fuel market, but in the intellectual climate of opinion on the use of nuclear power. The first programme is now coming to an end and a second is being planned. [11]

There have, inevitably, been criticisms [12] of a programme which has undergone such rapid fluctuations but these are largely political rather than technical. From an engineering point of view the nuclear power stations are achievements of a high order; not only have they created new skills and new industries but they have stimulated the development of many associated technologies, the results of which have been applied with considerable success in other established industries. An excellent description of the main engineering features of the British nuclear power stations is given in Hammond's book, [13] written mainly for the layman, which includes a review of the British nuclear power programme. A more detailed account [14] of the building of one of these stations has been written by the Public Relations Officer of the Berkeley Nuclear Power Station which was the first to come into operation in June 1962.

ECONOMIC ASPECTS OF NUCLEAR POWER

So long as there is a plentiful supply of conventional fuel, nuclear power stations will be built in large numbers only if they can be shown to be commercially competitive with other power sources. The economics of power reactor operation is still the subject of much discussion because, as yet, there is comparatively little operating experience on which to base calculations. This subject was discussed in a general way by Wendt [15] in 1957, and in more detail in *Atomic Power: an Appraisal,* [16] published in the same year, which is a record of informal panel discussions held in Washington at the eleventh annual general meeting of the Board of Directors of the World Bank. The chapter by the editor on "Economics of Nuclear Power" is a very lucid summary of the problem.

A more advanced treatment will be found in *Introduction to Nuclear Power Costs,* [17] which discusses the various methods of estimating the cost of each of the many factors which must be considered in measuring the overall cost of nuclear power, and shows how variations in the value of each affects the whole. One of the most important of these factors is, of course, the availability and market price of raw materials, and a recent assessment of this [18] has been made by the Euratom Supply Agency. This report estimates the output of natural and enriched uranium and plutonium in the West for the next five years and the European Community's requirements.

A factor which is not normally taken into account in estimating reactor costs is the cost of the research and development programme which is a necessary preliminary for any country wishing to build its own reactors. In the case of countries such as the U.S.A. and the U.K. this has been particularly high, an inevitable consequence of being first in the field. This question is examined in *Economics of Atomic Energy*, [19] in which the author considers the cost of an atomic energy programme in relation to the financial resources of a country and the impact of nuclear energy on the national economy, the implication being that there is a level of national income below which an atomic energy industry cannot be sustained.

NUCLEAR REACTORS

Most of the elementary introductions to atomic energy which were mentioned in chapter 7 devote considerable space to reactors without going into too much detail. A much fuller treatment is given in Jay's *Nuclear Power Today and Tomorrow* [20] which describes the main reactor types and discusses the principles and problems of nuclear power generation. This book contains a select bibliography and a glossary of technical terms and is an excellent introduction to the subject for those with or without scientific training. A similar introduction, written primarily for the layman, is *The New Power* [21] which is concerned mainly with power reactors but includes short chapters on other aspects of nuclear engineering.

A more elaborate introductory course for engineers and technicians planning to enter the atomic energy field is provided by *Nuclear Engineering Fundamentals* [22] which requires only a knowledge of algebra and elementary physics. The first three parts provide a background in the elements of atomic and nuclear physics; the last two deal mainly with reactors.

The literature on reactors is very extensive and reference should be made to the appropriate sections of the bibliographies mentioned in earlier chapters. More than 4000 of the most important articles on reactors, appearing between 1947 and 1959, are listed in the IAEA bibliography [23] on the subject. The references, arranged in ten broad categories, are taken from English, Russian, French, German, Italian and Japanese literature and have abstracts in either English or Russian, although all non-English titles are translated into English.

There are a number of excellent text books on nuclear engineering of which the most widely used is probably *Nuclear Reactor Engineering* [24] by Glasstone and Sesonske, which is sponsored by the USAEC and draws freely on material used at the Oak Ridge Reactor School. Intended as an introductory course for graduate engineering students the book covers every aspect of the subject with emphasis on the fundamental scientific and engineering principles of reactor systems and their design, and is well documented. A shorter and more elementary text is Murphy's *Elements of Nuclear Engineering* [25] which is a survey of the field for undergraduate students, written with the object of stimulating their interest in the subject. Two other introductory texts are those by Murray [26] and Stephenson, [27] both of which are written primarily for undergraduates following a conventional engineering course but intending to specialize in nuclear engineering.

The most comprehensive reference book on reactors is the *Reactor Handbook*, [28] sponsored by the USAEC and first issued as part of the *Selected Reference Material on Atomic Energy* (see Chapter 7). The four volumes represent the work of some hundreds of contributors and this is an essential reference work for technical personnel engaged in reactor research, design and development.

Nuclear reactors, whether designed for power production or research, fall into a number of basic types and the most widely used of these have received more detailed treatment in the literature. Separate volumes on pressurized water reactors, boiling water reactors, sodium graphite reactors and fluid fuel reactors have appeared in the *Atoms for Peace Series* (see Chapter 7, reference 57) as representing the principal types of reactors developed in the U.S.A. Gas-cooled, graphite-moderated reactors have been developed mainly in Britain and these are described in a book, [29] edited by Poulter, which is the work of a number of specialists drawn largely from the staff of the UKAEA. An introductory text on fast reactors, [30] by two Dounreay scientists, is also based primarily on British experience.

Power reactors, which may be of any of the types mentioned above, have received special attention in the literature. One of the most comprehensive and recent books is *Nuclear Power Plants* [31] which, apart from discussing principles and design, provides information on all nuclear power plants which have been built or are under construction throughout the world. The most significant references to documentary material on each reactor are given together with a list of code names of reactors. A more elementary text is Pearson's *Nuclear Power Technology* [32] which is based on lectures given at Birmingham College of Advanced Technology and offers a broad introduction to power reactor technology for engineers, managers and executives.

The information on power reactors made available at the two Geneva conferences in 1955 and 1958 is summarized in two books edited by Pickard. *Nuclear Power Reactors* [33] is concerned largely with pressurized water and boiling water reactors and derives mainly from American papers given at the 1955 conference. *Power Reactor Technology* [34] based on the 1958 conference includes material on gas-cooled reactors, fast reactors and heavy water reactors as well as those already mentioned. Both these volumes are intended for practising nuclear engineers and physicists, as is *Nuclear Power Engineering* [35] which deals with the fundamentals of thermal design and processes in reactors and power plants with emphasis on thermodynamics, heat transfer and fluid flow.

The majority of the texts mentioned so far are based on knowledge gained from experimental studies and from the design and operation of prototype reactors since actual operating experience with commercial nuclear power stations is necessarily limited. The position is, however, changing rapidly and the IAEA have recently published the proceedings of a conference [36] entitled *Operating Experience with Power Reactors* at which 44 papers were presented on experience with twenty power stations in different countries.

The construction of nuclear power stations has created a number of problems for the civil engineer, particularly in the design and manufacture of containment vessels. This aspect is considered in *Introduction to Structural Problems in Nuclear Reactor Engineering* [37] which is concerned with the design of pressure vessels and other nuclear structures and is basically a treatise on advanced structural analysis. Pressure vessels are also the subject of a conference [38] held at Glasgow in 1960, which dealt mainly with experience in the U.S.A. and the U.K., and of a recent textbook [39] for engineering graduates, which covers the elements of design of such structures.

REACTOR THEORY

The fission chain reaction in a reactor is a complicated process, the theoretical analysis of which is fundamental to the design of various types of reactor. Reactor theory is touched upon in many of the books already mentioned but reactor designers require a more detailed treatment and several books have been written to meet this need. The most comprehensive text at a fairly advanced level is that of Weinberg and Wigner, [40] which is a unified account of reactor theory and of the basic nuclear physics needed for an understanding of the subject, and is intended for practising physicists and engineers. *Elements of Nuclear Reactor Theory* [41] is based on lectures given at the Oak Ridge Reactor School and covers the basic physics and mathematics of reactor core design. Its British counterpart is *An Introduction to Reactor Physics*, [42] based on courses given at Harwell, which is concerned mainly with the physics of gas-cooled reactors but includes information on the physics of other types. Both these books are intended as introductions to the subject for graduate students in science and engineering as is the more recent book by Murray. [43]

Liverhart's *Elementary Introduction to Nuclear Reactor Physics* [44] is an undergraduate textbook for those taking degrees in nuclear science or engineering and is based on courses given at New York Maritime College, whereas Isbin's *Introductory Nuclear Reactor Theory* [45] extends beyond the elementary theory but falls short of advanced treatments which require a more thorough grounding in mathematics and quantum mechanics. At a more elementary level still is the *Textbook of Reactor Physics*, [46] which is written for the non-specialist with a limited knowledge of mathematics, and forms a useful introduction to the subject for technicians concerned with reactor design.

Reactor physics is very largely the study of neutron populations under various conditions and an essential compilation for the reactor physicist is *Neutron Cross Sections,* [47] originally issued as a Brookhaven report, and sometimes known as the "barn book" from the name of the unit in which nuclear cross sections are measured. This is a collection of graphs and tables of neutron cross-section data mainly in the range appropriate for reactor neutrons at thermal energies. To assist in the understanding and use of this compilation the first author has written a monograph [48] on the subject which gives a clear and concise explanation of the principles of cross section theory for neutrons of all energy ranges.

The data given in BNL—325 is applicable mainly to thermal reactors, i.e. the vast majority of the reactors in existence. Some data relating to fast reactor cross sections is given in a short monograph [49] by Yiftah and others published in 1960.

REACTOR CONTROL AND INSTRUMENTATION

A comprehensive compilation and analysis of existing information on the theory and design of reactor control systems is *Nuclear Reactor Control Engineering,* [50] which includes a bibliography of original papers on the subject and is written primarily for control engineers. An earlier book by Schultz [51] emphasizes the use of servomechanisms in the control of reactors, but is based mainly on experience with aqueous reactors such as the pressurized water type. Fozard's book [52] reflects British experience with gas-cooled reactors and is a practical textbook for physicists and engineers with little experience or advanced training in electrical control. *Nuclear Reactor Control and Instrumentation* [53] is an introduction to the subject for qualified technicians, research assistants and technical college students and appears in the Nuclear Engineering Monograph series published in association with the journal *Nuclear Engineering.*

REACTOR CATALOGUES

The most comprehensive list of reactors in operation or under construction, throughout the world, is that published by the IAEA in five volumes, [54] which covers power, research, test and experimental reactors in all the major nuclear countries. A more up-to-date list of American reactors is contained in the USAEC publication TID—8200 [55] which is revised every six months and covers civil and military reactors and critical assemblies. Both these directories give details of location, type, power output, purpose, main contractor, and start-up date (estimated in the case of those under construction). Reactors situated in Euratom countries are listed in a recent Euratom publication [56] which also covers uranium and thorium mines, ore concentrating and refining plants, uranium enrichment plants, fuel processing plants and waste processing installations in the countries of the Community. The main parameters of the world's power reactors are listed in an up-to-date review [57] produced by the Hanford Laboratories which includes, for each reactor, reference to more detailed descriptions in the literature.

NUCLEAR PROPULSION

Nuclear propulsion is merely another aspect of nuclear power technology but tends to be treated separately in the literature in common with other forms of motive power.

There is a considerable body of literature on this subject and the IAEA has issued a bibliography [58] of journal articles on the nuclear propulsion of aircraft, ships, rockets and other vehicles, covering the period 1950 to 1960. Abstracts are grouped under five broad subject headings and there is an author index. A bibliography on ship propulsion only, [59] covering roughly the same period, has been issued by the UKAEA and a select bibliography [60] of the unclassified literature on nuclear rockets has been published by the Office of Technical Services.

A number of textbooks and monographs are now available. *Nuclear Propulsion,* [61] edited by Thring, studies the various types of reactor suitable for propulsion systems and discusses ship propulsion and rocket propulsion in some detail. The level of the contributions varies considerably, some approaching the subject at a rather elementary level, others at a more advanced level. Ship propulsion is treated more fully in Kramer's book [62] which is sponsored by the USAEC as a source book on the nuclear propulsion of merchant vessels for shipping executives, port authorities, and construction and design engineers and may be read with a fair amount of ease by interested laymen. Much of the book is devoted to a description of the N. S. Savannah, the first nuclear merchant ship. Crouch's book [63] deals mainly with merchant ships and is an introductory text for marine engineers and others with little background in nuclear science. The subject is treated at a more advanced level in the papers presented at an IAEA symposium, the proceedings of which appeared in 1961, [64] although the main emphasis of the symposium was on the safety aspects of nuclear marine propulsion.

Nuclear rocket propulsion is still in the experimental stage and the book [65] by Bussard and De Lauer is based on research carried out at Los Alamos and is written mainly for research and development engineers.

REACTOR SAFETY

A chain reaction can take place in any lump of fissionable material once the critical size is exceeded and it is therefore necessary to take stringent precautions to ensure that amounts of fissile isotopes are always well below the critical size during storage, transport and processing. Only in the reactor, under controlled conditions, is the critical size exceeded and here it is necessary to design control systems in such a way that failure automatically renders the reactor safe, and to ensure that any abnormal release of radioactivity is contained. Even during normal operation the core of the reactor must be completely shielded so that operating personnel are fully protected against ionizing radiation.

All these aspects are dealt with in *Reactor Safeguards* [66] which is based on the author's experience while serving as secretary of the U.S. Reactor Safeguards Committee, a body which had the unenviable job of drafting recommendations, relating to reactor operation and construction, which ensured public safety yet permitted adequate progress to be made in the development of reactors.

The problems of criticality are explored in the *Nuclear Safety Guide* [67] which sets out to answer the question, "How can a chain reaction be prevented in fissionable materials being processed, stored or transported on an industrial scale?" The *Guide* considers the physics, engineering and administrative aspects of the problem and contains a list of references and a selected reading list. Recommendations [68] on the safe operation of reactors have been made by the IAEA, the recommendations being based on the best practice in experienced countries. The last point, that of protection against radiation, is the subject of McCullough's book [69] which summarizes papers given at the first Geneva Conference in 1955 and includes an account of safety criteria for reactors and radiochemical plant. A more comprehensive account of the theory and practice of reactor shielding will be found in *Radiation Shielding*, [70] a basic text book for physicists and engineers, which also deals with the shielding of particle accelerators. *Nuclear Reactor Shielding* [71] is an introduction to the problem of shielding, designed mainly for engineers and technicians, whereas Komarovskii's book [72] describes the use and properties of shielding materials, particularly concretes, and is an advanced text for design engineers.

There are, of course, a large number of articles and reports on reactor safety and reference should be made to Smith's bibliography [73] which covers material issued up to 1960.

CHEMICAL PROCESSING AND REACTOR MATERIALS

Uranium-235 is the only naturally occuring nuclear fuel but two others, plutonium-239 and uranium-233, can be made by the neutron irradiation of uranium-238 and thorium-232 respectively, the last two being known as fertile materials. Fuels may be used as liquids or solids, in the latter case in the form of fuel elements in which other substances, including fertile materials, may be present.

The fabrication of fuel elements is treated in some detail in the book edited by Kaufmann [74] which also contains papers on the properties of uranium, plutonium and thorium, their use as fuels and the processing of irradiated fuel elements to recover the fissile materials. This is an advanced text for metallurgists and reactor design engineers. *Chemical Processing of Nuclear Fuels*, [75] on the other hand, is an introduction for engineering graduates to the processing of fuels after irradiation, the emphasis being on basic principles rather than detailed technology. There is a large number of reports and articles on fuel processing and a bibliography [76] of the literature, in which most of the references are to American documents, has been published by OTS.

The placing of fuel elements in a reactor and their subsequent removal for processing, when they are highly radioactive, is a complex problem in materials handling and has received fairly comprehensive treatment in *Nuclear Fuel Handling*. [77] The author discusses the design and operation of fuel handling equipment and surveys existing and proposed installations in a number of countries.

Before fuel elements can be fabricated the raw materials, i.e. uranium and thorium, have to be obtained in a suitable form by extraction from ores and subsequent purification. These processes are described broadly in a book [78] by Grainger, written mainly for scientists and technologists who are not directly concerned with atomic energy. The book is authoritative and based on considerable experience in the field, yet is not beyond the intelligent layman. At a slightly more advanced level is *Chemical Processing of Reactor Fuels*, [79] which deals in some detail with solvent extraction processes and, although intended for nuclear engineering students, can be used as a reference book for those in the chemical processing field.

The basic material, of course, is uranium which, from being a little used curiosity, has become one of the most written-about metals in the world, particularly in the report literature. This is well demonstrated in the book by Gittus [80] which is a review of the world literature of uranium and covers every aspect of the subject from exploration for and mining of uranium ores to the effects of irradiation on the metal. It also includes a useful list of the world's reactors. Processing of uranium ores is the subject of the Atoms for Peace volume [81] by Clegg and Foley, based mainly on American practice. The next stage, of producing and fabricating high purity uranium metal is covered in *Uranium Production Technology* [82] which also reflects American practice. A comprehensive review of the metallurgy of uranium is the book by Wilkinson [83] which covers purification, fabrication, powder metallurgy, corrosion and uranium alloys and is well documented. The basic physical, chemical and mechanical properties of the metal are discussed in another Atoms for Peace volume by Holden [84] which includes material on the design of fuel elements and the effects of radiation damage.

Thorium, the other naturally occurring material, is the subject of *Thorium Production Technology*, [85] a source book for engineers and students covering all aspects of the subject. The metallurgy of thorium and its alloys is treated in *The Metal Thorium* [86] which includes information on non-nuclear applications of the metal.

The literature of the man-made fissile materials is not quite so extensive as that of uranium, but is growing rapidly. Wilkinson's book [87] includes an annotated bibliography of 45 pages covering the literature of plutonium up to 1959. This volume is based on a symposium held in February 1959 and deals with the recovery of plutonium from its salts and the physical metallurgy of the metal. A second plutonium conference held in April 1960 included papers [88] on the use of plutonium as a nuclear fuel as well as other aspects. *The Metal Plutonium* [89] is a review by a number of contributors of plutonium metallurgy and its use in reactors, and includes information made available at the 1958 Geneva conference. A more recent work by Taube [90] gives a comprehensive account of the chemistry, technology and nuclear properties of plutonium and its role in nuclear power.

One other aspect of fuel technology which has received some attention is the processing of enriched uranium. This leaves the gaseous diffusion plant in the form of uranium hexafluoride which has to be processed to obtain the enriched metal and this subject is covered in *Enriched Uranium Processing*, [91] which includes chapters on safety aspects and plant design and is well supplied with references.

There are a number of other materials of interest in nuclear engineering particularly graphite, beryllium, zirconium, magnesium, niobium, liquid metals and heavy water. *Chemical Processing in the Atomic Energy Industry* [92] is an introduction to the extraction and purification of some of these materials together with uranium, plutonium and thorium, and includes chapters on fuel processing and isotope separation. An earlier book on reactor materials [93] includes structural materials, fuels, ceramics and liquid metals and has a useful bibliography. *Materials for Nuclear Engineers* [94] is a compilation of data for use in design problems with adequate theoretical discussion to assist in extrapolation or interpolation. Based mainly on UKAEA experience it covers most of the materials already mentioned. Another practical handbook for engineers is Kopelmann's book [95] which describes the preparation, properties and behaviour of all the important reactor materials including moderators and coolants.

Graphite has been given separate treatment in *Nuclear Graphite* [96] which covers the physical, mechanical and thermal properties of graphite, the preparation of nuclear-grade graphite, the effects of irradiation on its properties and its use in reactors.

The main reference book on liquid metals is the *Liquid Metals Handbook*, [97] sponsored by the USAEC, which contains data on the physical and chemical properties of liquid metals and their use as a heat-transfer medium. A supplement [98] deals with the technology of sodium and sodium–potassium heat transfer systems in reactor engineering.

Most of the books mentioned in this section assume some knowledge of the basic chemistry of the materials discussed. *Chemistry in Nuclear Technology* [99] has been written for those who lack this grounding and covers radiochemistry, the chemistry of uranium, thorium and their compounds and the transuranium elements, processing of ores and irradiated fuels, waste disposal, isotope separation, and the chemistry of other reactor materials, and is designed as a means of enabling specialists in nuclear technology to absorb the disciplines basic to their field.

DISPOSAL OF RADIOACTIVE WASTES

Residual waste from reactors, fuel processing plants and radiochemical laboratories is highly radioactive and the disposal of such waste presents a problem of some complexity since much of this radioactivity is long-lived. *Radioactive Wastes,* [100] edited by Collins, contains a number of contributions by various authors covering all aspects of the treatment and disposal of active wastes and includes a chapter on the British legislation relating to the subject. *Atomic Energy Waste* [101] covers much the same field but devotes more space to the recovery of useful radioisotopes from fission products and their uses in research and agriculture. *The Treatment and Disposal of Radioactive Wastes* by Amphlett [102] emphasizes the chemical processing aspects and describes developments in waste disposal in atomic energy installations throughout the world. Some attention is given to the disposal of waste on a small scale, i.e. from radiochemical laboratories and similar institutions, where simple methods suffice.

The disposal of active waste has been the subject of a number of conferences. An international conference [103] held in 1959, and sponsored by the IAEA, UNESCO and the FAO, produced a number of original papers on different aspects of the subject. A later IAEA conference [104] considered the problem of high-activity wastes, such as those resulting from the processing of irradiated fuels, which are the most difficult to deal with. Most of the papers discussed methods for the solidification and fixation of liquid wastes, this being one of the most promising techniques in the field.

Low-activity liquid wastes can be conveniently discharged by pipe-line into the sea provided certain precautions are taken. A group of experts, set up by the IAEA in 1958, has studied the problem of preventing pollution of coastal waters by discharges of radioactive material, and published a report [105] and recommendations. The report also discusses the discharge of waste from nuclear-powered vessels in the deep sea. The disposal of waste into marine and fresh waters has received a good deal of attention in the literature and an IAEA bibliography [106] on the subject has over 2000 references covering the period 1955–1960. Selected reports and journal articles on the general problem of waste disposal are listed in a series of bibliographies published by the Office of Technical Services. [107]

The disposal of radioactive waste material, the cinders of the nuclear fire, is a problem to which a complete solution has not yet been found and since it touches upon the health and safety of the community it is essential that the layman understands something of the issues involved. A useful guide is *Disposal of Radioactive Wastes,* [108] by Saddington and Templeton, which introduces the problem to the general reader and outlines the methods so far employed for its solution. The book is based mainly on British practice and is a useful guide for those handling radioactive materials on a small scale.

Since this is essentially a public health problem legislation has been necessary to ensure that the public are adequately protected. In the United Kingdom the Government invited the Radioactive Substances Advisory Committee to set up a panel of experts, in 1956, to advise on the best methods of disposing of radioactive waste and, if legislation were necessary, to suggest the form it should take. The Panel issued a report which was published as a white paper [109] and is a very readable and concise statement of all aspects of the problem as it appears in Britain. The recommendations in the white paper formed the basis of the *Radioactive Substances Act, 1960* [110] which gave adequate powers to the Minister of Housing and Local Government to make whatever regulations were necessary to protect the public against this hazard.

REFERENCES

Energy Resources and Power Programmes

1. PUTNAM, P. C. *Energy in the future.* London, Macmillan, 1954.
2. THIRRING, H. *Power production: a practical application of world energy.* London, Harrap, 1956.
3. UNITED STATES ATOMIC ENERGY COMMISSION. *Fossil fuels in the future,* by M. F. Searl. Washington, Office of Technical Services, 1960. (TID—8209.)
4. LANSDELL, N. *The atom and the energy revolution.* Harmondsworth, Penguin Books, 1958.
5. UNITED STATES ATOMIC ENERGY COMMISSION. *Civilian nuclear power: appendices to "A report to the President—1962".* USAEC, Division of Technical Information Extension, Oak Ridge, Tennessee, 1963.
6. UNITED STATES ATOMIC ENERGY COMMISSION. *Power reactor development programs.* USAEC, Division of Technical Information Extension, Oak Ridge, Tennessee, 1963.
7. MULLENBACH, P. *Civilian nuclear power: economic issues and policy formation.* New York, Twentieth Century Fund, 1963.
8. LORD PRESIDENT OF THE COUNCIL AND MINISTER OF FUEL AND POWER. *A programme of nuclear power.* London, H.M.S.O., 1955. (Cmd. 9389.)
9. SECRETARY OF STATE FOR SCOTLAND AND MINISTER OF POWER. *Capital investment in the coal, gas and electricity industries. Appendix: The revised nuclear power programme.* London, H.M.S.O., 1957. (Cmnd. 132.)
10. MINISTER OF POWER. *The nuclear power programme.* London, H.M.S.O., 1960. (Cmnd. 1083.)
11. MINISTER OF POWER. *The second nuclear power programme.* London, H.M.S.O., 1964. (Cmnd. 2335.)
12. See, for example, GOLDRING, M. Replanning Britain's nuclear power programme. *In: J. Brit. Nucl. Energy Soc.,* vol. 3, no. 1, pp. 11–15, January 1964.

13. HAMMOND, R. *British nuclear power stations.* London, Macdonald, 1961.
14. HALL, J. *The story of the construction of Berkeley Nuclear Power Station.* London, L. Hill, 1963.

Economic Aspects of Nuclear Power

15. WENDT, G. *Prospects of nuclear power and technology.* London, Macmillan, 1957.
16. ALLARDICE, C., ED. *Atomic power: an appraisal.* London, Pergamon Press, 1957.
17. ROCHMAN, A. *Introduction to nuclear power costs.* New York, Simmons-Boardman Pub. Co., 1959.
18. EURATOM SUPPLY AGENCY. *State of the nuclear fuel market in the Community and the free world in the period 1963—1967.* Brussels, Euratom, 1963. (Circular no. 4.)
19. GOLDRING, M. *Economics of atomic energy.* London, Butterworth, 1957.

Nuclear Reactors

20. JAY, K. E. B. *Nuclear power today and tomorrow.* London, Methuen, 1961.
21. LONG, G., PRICE, D. AND SOWDEN. R. G. *The new power.* London, Newnes, 1962.
22. WEINSTEIN, R., BOLTAX, A. AND LANZA, G. *Nuclear engineering fundamentals.* New York, London, McGraw-Hill, 1964. 5 vols in 1.
    Vol. 1: Atomic physics; Vol. 2: Nuclear physics; Vol. 3: Interaction of radiation with matter; Vol. 4: Nuclear materials; Vol. 5: Nuclear reactor theory.
23. INTERNATIONAL ATOMIC ENERGY AGENCY. *Nuclear reactors.* Vienna, IAEA, 1960. (Bibliographical series no. 2.)
24. GLASSTONE, S. AND SESONSKE, A. *Nuclear reactor engineering: prepared under the auspices of the Division of Technical Information, USAEC.* Princeton, Van Nostrand, 1963.
25. MURPHY, G. *Elements of nuclear engineering.* New York, Wiley, 1961.
26. MURRAY, R. L. *Introduction to nuclear engineering.* 2nd ed. New Jersey, Prentice-Hall, 1961.
27. STEPHENSON, R. *Introduction to nuclear engineering.* 2nd ed. New York, McGraw-Hill, 1958.
28. *Reactor handbook.* 2nd ed. New York, Interscience, 4 vols. (Sponsored by USAEC.)
    Vol. 1: Materials, 1960; Vol. 2: Fuel reprocessing, 1961; Vol. 3: Physics and shielding, part A, Physics, 1962; part B, Shielding, 1962; Vol. 4: Engineering, 1963.
29. POULTER, D. R., ED. *Design of gas-cooled, graphite-moderated reactors.* London, O.U.P., 1963.
30. PALMER, R. G. AND PLATT, A. *Fast reactors.* London, Temple Press, 1961. (Nuclear Engineering Monographs.)
31. LOFTNESS, R. L. *Nuclear power plants: design, operating experience and economics.* Princeton, Van Nostrand, 1964.
32. PEARSON, F. J., ED. *Nuclear power technology.* London, O.U.P., 1963.
33. PICKARD, J. K., ED. *Nuclear power reactors.* Princeton, Van Nostrand, 1957.
34. PICKARD, J. K. AND OTHERS, EDS. *Power reactor technology.* Princeton, Van Nostrand, 1961.
35. EL-WABIL, M. M. *Nuclear power engineering.* New York, McGraw-Hill, 1962.
36. INTERNATIONAL ATOMIC ENERGY AGENCY. *Operating experience with power reactors: proceedings of an IAEA conference held in Vienna, June, 1963.* Vienna, IAEA, 1964. 2 vols. (28 papers in English, 14 in French, 2 in Russian.)
37. RYDZEWSKI, J. R., ED. *Introduction to structural problems in nuclear reactor engineering.* Oxford, Pergamon Press, 1962.

38. GLASGOW. ROYAL COLLEGE OF SCIENCE AND TECHNOLOGY. *Nuclear reactor containment buildings and pressure vessels: proceedings of a symposium, organised by the Department of Mechanical, Civil and Chemical Engineering, May, 1960.* London, Butterworth, 1960.
39. HARVEY, J. F. *Pressure vessel design: nuclear and chemical applications.* Princeton, Van Nostrand, 1963.

Reactor Theory

40. WEINBERG, A. M. AND WIGNER, E. P. *The physical theory of neutron chain reactors.* Chicago, University of Chicago Press, 1958.
41. GLASSTONE, S. AND EDLUND, M. C. *Elements of nuclear reactor theory.* Princeton, Van Nostrand, 1952.
42. LITTLER. D. J. AND RAFFLE, J. F. *An introduction to reactor physics.* 2nd ed. Oxford, Pergamon Press, 1957.
43. MURRAY, P. L. *Nuclear reactor physics.* London, Macmillan, 1959.
44. LIVERHART, S. E. *Elementary introduction to nuclear reactor physics.* New York, Wiley, 1960.
45. ISBIN, H. S. *Introductory nuclear reactor theory.* New York, Reinhold; London, Chapman and Hall, 1963.
46. HILL, J. F. *Textbook of reactor physics: an introduction prepared under the auspices of the UKAEA.* London, Allen and Unwin, 1961.
47. HUGHES, D. J. AND SCHWARTZ, R. B. *Neutron cross sections.* Washington, U.S. Government Printing Office, 1958. (Report no. BNL—325, 2nd ed.) Supplement no. 1, 1960.
48. HUGHES, D. J. *Neutron cross sections.* London, Pergamon Press, 1957.
49. YIFTAH, S. AND OTHERS. *Fast reactor cross sections.* Oxford, Pergamon Press, 1960.

Reactor Control and Instrumentation

50. HARRER, J. M. *Nuclear reactor control engineering.* Princeton, Van Nostrand, 1963.
51. SCHULTS, M. A. *Control of nuclear reactor and power plants.* 2nd ed. New York, McGraw-Hill, 1961.
52. FOZARD, B. *Instrumentation and control of nuclear reactors.* London, Iliffe, 1963.
53. BOWEN, J. H. AND MASTERS, E. F. O. *Nuclear reactor control and instrumentation.* London, Temple Press, 1959. (Nuclear Engineering monograph.)

Reactor Catalogues

54. INTERNATIONAL ATOMIC ENERGY AGENCY. *Directory of nuclear reactors.* (In English only.) Vienna, IAEA, 5 vols.
Vol. 1: Power reactors, 1959. (Out of print: superseded by vol. 4.); Vol. 2: Research, test and experimental reactors, 1959. (Information on 77 reactors in 17 countries.); Vol. 3: Research, test and experimental reactors, 1960. (Supplement to vol. 2. Information on 96 reactors in 20 countries.); Vol. 4: Power reactors, 1962. (Revised edition of vol. 1. Information on 56 reactors in 11 countries.); Vol. 5: Research, test and experimental reactors, 1964 (Supplement to vols. 2 and 3.)
55. UNITED STATES ATOMIC ENERGY COMMISSION. DIVISION OF REACTOR DEVELOPMENT. *Nuclear reactors built, being built, or planned in the United States as of June 30, 1963.* Washington, Office of Technical Services, 1963. (TID—8200, 8th rev.)
56. EURATOM. *Nuclear installations in the countries of the European Atomic Energy Community.* 2nd ed. Brussels, Euratom, 1963. (EUR—183 e.)

57. HANFORD ATOMIC PRODUCTS OPERATION, RICHLAND. *Review of power and heat reactor designs: domestic and foreign*, comp. by E. R. Appleby. Washington, Office of Technical Services, October, 1963. (HW—66666, rev. 2.)

Nuclear Propulsion

58. INTERNATIONAL ATOMIC ENERGY AGENCY. *Nuclear propulsion.* Vienna, IAEA, 1961. (Bibliographical series no. 3.)
59. YATES, B., COMP. *Bibliography on nuclear propulsion for ships.* London, H.M.S.O., 1960. (DEG—Inf. Ser. 242.)
60. BUSSARD, R. W., COMP. *Nuclear rocket propulsion: a selected bibliography of the unclassified literature.* Washington, Office of Technical Services, 1961. (LAMS—2519.)
61. THRING, M. W., ED. *Nuclear propulsion.* London, Butterworth, 1960.
62. KRAMER, A. W. *Nuclear propulsion for merchant ships: prepared under the auspices of the USAEC Divisions of Technical Information.* Washington, U.S. Government Printing Office, 1962.
63. CROUCH, H. F. *Nuclear ship propulsion.* Cambridge, Maryland, Cornell Maritime Press, 1960.
64. INTERNATIONAL ATOMIC ENERGY AGENCY. *Nuclear ship propulsion: proceedings of the symposium on nuclear ship propulsion with special reference to nuclear safety.* Vienna, IAEA, 1961.
65. BUSSARD, R. W. AND DE LAUER, R. D. *Nuclear rocket propulsion.* New York, McGraw-Hill, 1958.

Reactor Safety

66. RUSSEL, C. R. *Reactor safeguards.* Oxford, Pergamon Press, 1962.
67. AMERICAN STANDARDS ASSOCIATION, SECTIONAL COMMITTEE N 6 AND AMERICAN NUCLEAR SOCIETY, STANDARDS COMMITTEE. *Nuclear safety guide.* Washington, Office of Technical Services, 1961. (TID—7016, rev. 1.)
68. INTERNATIONAL ATOMIC ENERGY AGENCY. *Safe operation of critical assemblies and research reactors.* Vienna, IAEA, 1961. (Safety series no. 4.) Manual is available in English, French, Russian and Spanish editions.
69. MCCULLOUGH, C. R., ED. *Safety aspects of nuclear reactors.* Princeton, Van Nostrand, 1957.
70. PRICE, B. T., HORTON, C. C. AND SPINNEY, K. T. *Radiation shielding.* Oxford, Pergamon Press, 1957.
71. HARRISON, J. R. *Nuclear reactor shielding.* London, Temple Press, 1958. (Nuclear Engineering monographs.)
72. KOMAROVSKII, A. N. *Shielding materials for nuclear reactors,* translated from Russian by V. M. Newton. Oxford, Pergamon Press, 1961.
73. UNITED STATES ATOMIC ENERGY COMMISSION, OFFICE OF TECHNICAL INFORMATION. *Reactor safety: a literature search,* comp. by R. J. Smith. Washington, Office of Technical Services, 1960. (TID—3525, rev. 2.)

Chemical Processing and Reactor Materials

74. KAUFMANN, A. R., ED. *Nuclear reactor fuel elements; metallurgy and fabrication.* New York, Interscience, 1962.
75. MARTIN, F. S. AND MILES, G. L. *Chemical processing of nuclear fuels.* London, Butterworth, 1958.
76. (a) UNITED STATES ATOMIC ENERGY COMMISSION, OFFICE OF TECHNICAL INFORMATION. *Reprocessing of irradiated fission reactor fuel and breeding material: an annotated bibliography of selected report literature.* Washington, Office of Technical Services, 1958. (TID—3312.) Reports only up to 1958.

(b) UNITED STATES ATOMIC ENERGY COMMISSION, DIVISION OF TECHNICAL INFORMATION. *Reprocessing of irradiated fission reactor fuel and breeding material: an annotated bibliography of selected literature.* Washington, Office of Technical Services, June 1964. (TID–3312, suppl. 1.) Covers reports and articles 1958 to January 1964.

77. WORDSWORTH, A. D. *Nuclear fuel handling.* London, Butterworth, 1963.
78. GRAINGER, L. *Uranium and thorium.* London, Newnes, 1958.
79. FLAGG, J. F., ED. *Chemical processing of reactor fuels.* New York, Academic Press, 1961.
80. GITTUS, J. H. *Uranium.* London, Butterworth, 1963.
81. CLEGG, J. W. AND FOLEY, D. D., EDS. *Uranium ore processing.* Reading (Mass.), Addison-Wesley, 1958.
82. HARRINGTON, C. D. AND RUEHLE, A. E., EDS. *Uranium production technology.* Princeton, Van Nostrand, 1959.
83. WILKINSON, W. D. *Uranium metallurgy.* New York, London, Interscience, 1962. 2 vols.
   Vol. 1: Uranium process metallurgy; Vol. 2: Uranium corrosion and alloys.
84. HOLDEN, A. N. *Physical metallurgy of uranium.* Reading (Mass.), Addison-Wesley, 1958.
85. CUTHBERT, F. L. *Thorium production technology.* Reading (Mass.), Addison-Wesley, 1958.
86. WILHELM, H. A., ED. *The metal thorium.* Cleveland, American Society for Metals, 1958.
87. WILKINSON, W. D., ED. *Extractive and physical metallurgy of plutonium and its alloys: including an annotated bibliography by W. D. Wilkinson.* New York, London, Interscience, 1960.
88. SOCIÉTÉ FRANÇAISE DE METALLURGIE and COMMISSARIAT A L'ÉNERGIE ATOMIQUE. *Plutonium 1960; proceedings of the second International Conference on Plutonium Metallurgy, Grenoble, April, 1960.* London, Cleaver-Hume, 1961.
89. COFFINBERRY, A. S. AND MINER, W. N. EDS. *The metal plutonium.* Chicago, University of Chicago Press, 1961.
90. TAUBE, M. *Plutonium.* Oxford, Pergamon Press, 1963.
91. PATTON, F. S. AND OTHERS. *Enriched uranium processing.* Oxford, Pergamon Press, 1963.
92. COOPER, A. R. *Chemical processing in the atomic energy industry.* London, Iliffe, 1964.
93. FROST, B. R. T. AND WALDRON, M. B. *Nuclear reactor materials.* London, Temple Press, 1959. (Nuclear Engineering monographs.)
94. MCINTOSH, A. B. AND HEAL, T. J. EDS. *Materials for nuclear engineers.* London, Temple Press, 1960.
95. KOPELMANN, B., ED. *Materials for nuclear reactors.* New York, McGraw-Hill, 1959.
96. NIGHTINGALE, R. E., ED. *Nuclear graphite.* New York, Academic Press, 1962.
97. LYON, R. N., ED. *Liquid-metals handbook.* 2nd ed. Washington, USAEC, 1952. (Now available from University Microfilms Inc. as OP no. 4056.)
98. JACKSON, C. B., ED. *Liquid-metals handbook: sodium—NaK supplement.* Washington, USAEC, 1955. (Now available from University Microfilms Inc. as OP no. 10842.)
99. PETERSON, S. AND WYMER, R. G. *Chemistry in nuclear technology.* Oxford, Pergamon Press; Reading (Mass.), Addison-Wesley, 1963.

Disposal of Radioactive Waste

100. COLLINS, J. C., ED. *Radioactive wastes: their treatment and disposal.* London, Spon, 1960.

101. GLUECKAUF, E., ED. *Atomic energy waste; its nature, use and disposal.* London, Butterworth, 1961.
102. AMPHLETT, C. B. *Treatment and disposal of radioactive wastes.* Oxford, Pergamon Press, 1961.
103. INTERNATIONAL ATOMIC ENERGY AGENCY. *Disposal of radioactive wastes; proceedings of a conference on the disposal of radioactive waste, sponsored by the IAEA and UNESCO with the co-operation of the FAO, held in Monaco, November, 1959.* Vienna, IAEA, 1960. 2 vols.
104. INTERNATIONAL ATOMIC ENERGY AGENCY. *Treatment and storage of high level radioactive wastes: proceedings of the symposium ... held by the IAEA in Vienna, October, 1962.* Vienna, IAEA, 1963.
105. INTERNATIONAL ATOMIC ENERGY AGENCY. *Radioactive waste disposal into the sea.* Vienna, IAEA, 1961. (Safety series no. 5.)
106. INTERNATIONAL ATOMIC ENERGY AGENCY. *Disposal of radioactive wastes into marine and fresh waters.* Vienna, IAEA, 1962. (Bibliographical series no. 5.)
107. (a) UNITED STATES ATOMIC ENERGY COMMISSION, OFFICE OF TECHNICAL INFORMATION. *Radioactive waste processing and disposal: a bibliography of selected report literature.* Washington, Office of Technical Services, 1958. (TID–3311.)
Covers reports issued up to the end of 1957.
(b) UNITED STATES ATOMIC ENERGY COMMISSION, OFFICE OF TECHNICAL INFORMATION. *Radioactive waste processing and disposal: a literature search.* Washington, Office of Technical Services, 1960. (TID–3555.)
Covers reports issued January 1958—May 1960 and journal articles issued January 1951—May 1960.
(c) UNITED STATES ATOMIC ENERGY COMMISSION, DIVISION OF TECHNICAL INFORMATION. *Radioactive waste processing and disposal: a literature search.* Washington, Office of Technical Services, August 1962. (TID–3555, suppl. 1.)
Reports and articles issued June 1960—May 1962.
(d) UNITED STATES ATOMIC ENERGY COMMISSION, DIVISION OF TECHNICAL INFORMATION. *Radioactive waste processing and disposal: an annotated bibliography of selected literature.* Washington, Office of Technical Services, June 1964. (TID–3311, suppl. 1.)
Reports and articles issued June 1962—January 1964.
108. SADDINGTON, K. AND TEMPLETON, W. L. *Disposal of radioactive waste.* London, Newnes, 1958.
109. MINISTRY OF HOUSING AND LOCAL GOVERNMENT, AND OTHERS. *The control of radioactive wastes. (Report of the Panel on Disposal of Radioactive Wastes.)* London, H.M.S.O., 1959. (Cmnd. 884.)
110. GREAT BRITAIN, STATUTES. *Radioactive Substances Act, 1960.* London, H.M.S.O., 1960.

APPENDIX

PERIODICALS CONCERNED MAINLY WITH NUCLEAR ENGINEERING, TECHNOLOGY AND MATERIALS

1. *Advances in Nuclear Science and Technology.* Academic Press, 111 Fifth Ave., New York 3, N.Y. 1962 and onwards. (Annual.)
Authoritative and critical review articles covering all phases of atomic energy except theoretical physics and radiation biology and medicine, with the main emphasis on nuclear reactors and nuclear engineering.

2. *American Nuclear Society Transactions.* American Nuclear Society, 244 East Ogden Ave., Hinsdale, Illinois. 1958 and onwards. (Twice yearly.)
Programmes and summaries of papers presented at the annual meeting in June and the winter meeting in November of each year. The majority of papers relate to reactor technology and nuclear engineering.
3. *Bibliography of Geological Literature on Atomic Energy Raw Materials.* Geological Survey of Great Britain, Atomic Energy Division, Young St., Kensington, London, W.8. 1964 and onwards. (Monthly.)
Continuation of "Bibliography of Geological Literature bearing on Atomic Energy", 1950—1963. About 400 references p.a. on beryllium, thorium, uranium and other nuclear materials, radioactivity, geological dating and prospecting techniques.
4. *Journal of the British Nuclear Energy Society.* British Nuclear Energy Society, 1—7 Great George St., London, S.W.1. 1956 and onwards. (Quarterly.)
Original papers, including those presented at meetings of the Society, and notes of discussions on reactor engineering and nuclear technology.
5. *Journal of the Joint Panel on Nuclear Marine Propulsion.* Institute of Marine Engineers, Memorial Building, 76 Mark Lane, London, E.C.3. 1957 and onwards. (Irregular.)
Papers presented at meetings of the Panel, which is representative of the Institute of Marine Engineers, the Royal Institution of Naval Architects, the Institution of Engineers and Shipbuilders in Scotland, and the North East Coast Institution of Engineers and Shipbuilders.
6. *Journal of Nuclear Materials.* North Holland Publishing Co., P.O. Box 103, Amsterdam. 1959 and onwards. (Monthly.)
Articles on metallurgy, ceramic materials and solid state physics in relation to nuclear science. Articles in English, French or German with abstracts in English.
7. *Nuclear Engineering (incorporating Nuclear Power).* Temple Press, Bowling Green Lane, London, E.C.1. 1956 and onwards. (Monthly.)
Short articles and comment on developments in the nuclear power industry. Technical articles on reactor engineering and nuclear technology. In English with annotated summaries of contents in French and German.
8. *Nuclear Power Patents Bulletin.* Derwent Information Service, Rochdale House, Theobalds Rd., London, W.C.1. 1961 and onwards. (Monthly.)
Approximately 6000 references p.a. on British and foreign patents relating to the production and use of nuclear power and ionizing radiations.
9. *Nuclear Science and Engineering.* American Nuclear Society, Hinsdale, Illinois. 1956 and onwards. (Monthly.)
Original articles on reactor physics and engineering, many of them amplifying work previously summarized in the *Transactions*.
10. *Reactor Science and Technology (Journal of Nuclear Energy,* parts A/B). Pergamon Press, Headington Hill Hall, Oxford. Vol. 14. 1961 and onwards. (Monthly.)
Original articles on reactor physics and engineering including translations of selected articles from *Atomniya Energiya.* Previously issued as (1) *Journal of Nuclear Energy,* vols 1—9, 1954—1959, (2) *Journal of Nuclear Energy,* part A, Reactor Science, vols. 10—13, 1959—1961 *and* part B, Reactor Technology, vol. 1, 1959—1961.
11. *Research Reactor Journal.* AMF Atomics, 140 Greenwich Ave., Greenwich, Connecticut. 1960 and onwards. (Quarterly.)
Articles on the operation of, and experiments with, research reactors.

CHAPTER 10

IONIZING RADIATIONS AND RADIOISOTOPES

GENERAL

Radioisotopes and other radiation sources are used in industry, agriculture, medicine and research in a variety of ways although, basically, all applications can be grouped into three categories. First, tracer applications, which depend on the fact that a radioisotope is chemically identical with any other isotope of the same element but can be detected at any stage of a reaction or process by the penetrating radiation which is emitted. Secondly, since radiation is attenuated or absorbed by material in proportion to its thickness, density or nature, small radiation sources can be used to measure thickness, density or composition. Thirdly, radiation in more massive quantities effects changes in inorganic and organic substances including living organisms, thus modifying their composition or properties.

Radioisotopes are used in all three categories, but in the last category the source may be and frequently is a particle accelerator, a nuclear reactor or a pulsed neutron generator. Whatever the process or application being considered, a knowledge of the nature of radiation and of its effects on materials is required together with an understanding of methods of detecting and measuring radiation and of the hazards involved in working with radioactive materials.

All these aspects are considered in *Nuclear Radiation Physics,* [1] an introduction which requires only an elementary knowledge of mathematics and which is well furnished with selected references for further reading. For scientific workers who have no specialized knowledge of the subject, one of the best introductions is *Radioactive Isotopes,* [2] by Whitehouse and Putman, which gives a comprehensive account of basic principles, the properties, detection and measurement of radiation, and the production, use and safe handling of radioisotopes and contains a large number of references to original papers.

The majority of popular books on atomic energy, including those already mentioned in Chapter 7, have sections on isotopes and radiation. In addition there are a number of elementary introductions devoted entirely to these subjects. *The Uses and Effects of Nuclear Energy* [3] is a collection of essays on various aspects of the biological effects of radiation but includes a section on radioisotope applications in industry, medicine and agriculture. Applications in all fields is the theme of Putman's *Isotopes*, [4] probably the best general introduction for the layman and the beginner. *Labelled Atoms*, [5] by Glascock, and *Applied Atomic Energy*, [6] by Fearnside, Jones and Shaw, are both concerned mainly with biological applications of isotopes, whereas Jefferson's *Radioisotopes* [7] covers industrial applications and is based on experience with the UKAEA Isotope Advisory Service. Although of interest to the general reader its main purpose is to provide industrialists with an appreciation of the benefits to be derived from the use of isotopes in industry.

SEPARATION AND PRODUCTION OF ISOTOPES

One of the earliest instruments for separating isotopes on a laboratory scale was the mass spectrometer (see Chapter 8, references 28–31) and this method is described by Aston in his *Mass Spectra and Isotopes*, [8] which contains a review of all the chemical elements and their naturally occurring isotopes, with references to the original papers describing their isolation. Since Aston's day electromagnetic separation has developed rapidly and is now used for the production of isotopes, particularly stable isotopes, on an industrial scale as well as in the laboratory. Two international conferences have been held on this subject. The first, [9] at Harwell, was concerned mainly with the design of equipment for electromagnetic separation of stable isotopes; the second, [10] at Vienna, was exclusively concerned with the separation of radioactive isotopes. Information up to 1958 has been consolidated in *Electromagnetic Isotope Separators*, [11] an introduction for the potential worker in this field who is already familiar with the principles of ion optics and mass spectroscopy.

Many other methods of separation have been used on both laboratory and industrial scales and the most promising of these were discussed at an international conference held in Amsterdam in 1957. [12] Current knowledge in this field is summarized comprehensively in *Separation of Isotopes* [13] which discusses the theory and practice of the commonly used methods of separation in the laboratory and on an industrial scale, the contributors all being experts in their respective fields.

Artificially produced radioisotopes are made by subjecting elements and compounds to intense radiation, either in a reactor or by using a particle accelerator. Some of the isotopes made in this way have extremely short half-lives and the problems associated with producing these have been discussed at a recent international conference [14] sponsored by the IAEA.

APPLICATIONS OF ISOTOPES AND RADIATION

A fairly comprehensive guide to sources of American information in this field has been issued by the USAEC, [15] the bibliography listing over 600 references to published information and to services. Abstracts of publications in languages other than English appear in a continuing bibliography, [16] also issued by the USAEC, which covers the period from 1959 onwards, with indexes to authors and country of origin. Tracer techniques are the subject of a recent bibliography, [17] compiled by the UKAEA Radiochemical Centre, which includes selected references to publications on the preparation of labelled compounds, applications in all fields, radiation measurement and safety precautions.

There have been four major international conferences on techniques for using radioisotopes and the results obtained in various fields. In the first three, held at Oxford [18, 19] and Paris, [20] applications in research, industry, medicine and biology were reported, but by the time of the fourth conference in Copenhagen [21] the variety of applications had grown to such an extent that only those in the physical sciences and industry were considered; even so the proceedings ran to three volumes.

APPLICATIONS IN THE PHYSICAL SCIENCES AND INDUSTRY

The extent to which radioisotopes are used in industry and physical research is not known precisely but some quantitative estimates have been made. A report [22] compiled by the Illinois Institute of Technology in 1962 describes a comparative study of radioisotope use in various countries, with emphasis on the U.S.A. and the U.S.S.R. Estimates of the extent to which isotopes are used in various fields are based on a study of the literature of the subject and particularly of conference proceedings. A more accurate estimate of use in Great Britain can be obtained from a report to be published by the Social Survey Division of the Central Office of Information, a summary of which appeared in *Atom* [23] in August 1964. Apart from statistics of use the report gives some estimate of the economic benefits gained from using radioisotopes and discusses the growth of isotope techniques in Britain.

A detailed account of American developments in the application of isotopes in research and industry is given in a USAEC report [24] which summarizes earlier research and surveys carried out under the Commission's Isotope Development Program.

On the international scale, the IAEA has recently made a survey [25] of radioisotope applications classified by industry or economic activity, with selected references to the literature, grouped under the headings of Mining, Manufacturing, Building and Public Utility Services.

A brief introduction, for the industrialist and general reader, to some of the more common industrial applications is *Radioisotopes for Industry*, [26] by Rochlin and Schultz, and a comprehensive reference book for those using isotopes in chemistry, physics and engineering is *Radioisotope Applications Engineering*, [27] by Kohl and others.

One of the main uses of isotopes in industry is in the field of instrumentation and control, a subject covered in a recent Russian publication, [28] now available in translation. The general principles of control using isotopes and some specific problems relating to density and thickness gauges, level indicators, flow measurement and the control of composition are discussed. The most comprehensive text on this subject, however, which also considers radiation sources for industrial processing, is, as yet, available only in the German original. [29] A survey of the use of isotopes in measuring thickness and density appeared in two articles in *Atom* [30] in 1961 and techniques for using isotopes and other radiation sources in industrial radiography are described in Rockley's *Introduction to Industrial Radiology,* [31] a standard work in this field.

Tracer techniques used in chemical analysis and industrial processes are described in Daudel's book, [32] a straightforward and well-documented account suitable for anyone with a little scientific background. The technique of isotope dilution analysis is outlined in a booklet, [33] issued by the UKAEA Radiochemical Centre, which contains a useful bibliography. A more detailed account which includes other analytical applications is Lambie's manual, [34] designed for the research worker who has little previous knowledge of radiochemistry.

Another analytical technique is that of radioactivation analysis in which the substances to be analysed are activated in a nuclear reactor or by means of some other source of radiation. *Neutron Irradiation and Activation Analysis* [35] introduces the principles of this technique with some examples of applications. A fuller treatment is found in *Radioactivation Analysis* [36] which includes a survey of applications in geochemistry, biology and inorganic analysis.

Applications in Biology, Medicine and Agriculture

Radioactive isotopes are used in biological research as tracers and as radiation sources to determine the effects of radiations on biological organisms. In medicine they are used as tracers in diagnosis and as sources of intense radiation in radiotherapy. A short survey [37] of research carried out in biology and biochemistry with the aid of isotopes has been published by the IAEA; based on work published in 1959 it has nearly 600 references to original articles and monographs.

One of the earliest textbooks on tracer techniques is Hevesy's *Radioactive Indicators,* [38] which describes methods of using isotopes as tracers in biological and medical studies, and discusses the results obtained. Another book on methodology is Comar's *Radioisotopes in Biology and Agriculture,* [39] in which the emphasis is on the practical aspects. The bulk of the book is concerned with the properties of the most important individual isotopes and the procedures for using each of them and enough information is given to enable the reader to determine the appropriateness of a particular isotope for a given experiment. A rather similar approach is followed in Kamen's *Isotopic Tracers in Biology* [40] which covers stable as well as radioactive isotopes. Probably the most useful practical manual is *Isotopic Tracers,* [41] by Francis and others, based on a course given at the Medical College of St. Bartholomew's Hospital, London, which includes selected references at the end of each chapter.

Tracer applications in biochemistry are discussed by Sacks [42] in a book which emphasizes the broad principles and results obtained rather than detailed experimental procedures. The latter are described by Broda [43] who also discusses the use of isotopes as radiation sources in biochemical research.

A comprehensive text on medical applications is the three-volume work *Radioactive Isotopes in Medicine and Biology* [44] of which the first volume, dealing with basic principles, is suitable for isotope users in any field, and the second is designed for the physician with little previous training in radiochemistry. The third volume, which has not yet appeared, will cover biological applications. *Radioisotope Techniques in Clinical Research*, [45] by Veall and Vetter, and *Clinical Use of Radioisotopes*, [46] by Beierwaltes and others, are devoted mainly to techniques for using isotopes in diagnosis and more than half of the latter book is concerned with a single isotope, iodine-131.

The use of isotopes in diagnosis, radiotherapy and general medical research is the subject of Halnan's *Atomic Energy in Medicine*, [47] which is a semipopular account, and the *Manual for Nuclear Medicine*, [48] by King and Mitchell, which is designed for practitioners and students and is well documented. *Clinical Use of Radioisotopes*, [49] edited by Fields and Seed, deals with the clinical aspects of diagnosis and therapy and devotes some space to the planning and operation of the medical radioisotope laboratory. The use of radiation for the treatment of certain diseases is becoming increasingly general and there is a considerable volume of literature. General aspects are discussed by Delario [50] in a comprehensive account of the diseases which can be treated in this way and of the techniques used and by Paterson [51] in a more recent work.

A bibliography [52] of selected references to the literature published up to 1958 has been issued by the USAEC, most of the items relating to journal articles. The bibliography covers diagnosis, therapy, clinical research, physiological research, biochemistry and general medical research. Another useful reference source in this field is Etter's *Glossary* [53] which covers conventional radiological techniques and nuclear medicine.

In addition to the book by Comar [39] already mentioned, many of the general books on biological applications of isotopes include material on agricultural uses. Dick's *Atomic Energy in Agriculture* [54] and Singleton's *Nuclear Radiation in Food and Agriculture* [55] are both based on information presented at the first Geneva Conference in 1955, although the former is presented in a narrative form suitable for the general reader. A more up-to-date review [56] of the subject for the research scientist has been published by the IAEA, with selected references to over 120 original papers, and a more comprehensive and continuing bibliography [57] has been issued by the USAEC.

DIRECTORIES AND CATALOGUES OF ISOTOPES

Information about radioisotopes and labelled compounds available from the major world suppliers is given in the IAEA's *International Directory of Radioisotopes*, [58] which lists over 60 suppliers, and for each isotope gives details of the chemical form in which it is made available, the supplier, unit size and weight, specific and total activity, and price. Radioisotopes produced and sold by the UKAEA are listed in the catalogue [59] published by the

Radiochemical Centre which also includes information on the irradiation services which are available. To assist users in selecting the most suitable radioactive substances for their purpose the Radiochemical Centre publishes the *Radiochemical Manual*, [60] in two parts. The first part is a compilation of physical data relating to the isotopes in common use, with details of production processes and the physical and chemical forms in which isotopes are offered: the second part discusses generally the chemical characteristics of radioisotopes which must be taken into account when experiments are planned and describes, more specifically, the properties of the isotopes of the six most commonly used elements and their labelled compounds.

Stable and radioactive isotopes and labelled compounds sold commercially in the U.S.A. are listed in the *Isotope Index* [61] which is revised annually. In addition the Oak Ridge National Laboratory, the principal source of isotopes in the U.S.A., issues separate catalogues for radioactive and stable isotopes. [62]

ISOTOPE LABORATORY TECHNIQUES

The widespread use of radioisotopes has led to an increasing need for information on handling techniques, particularly from researchers and technicians who have had no previous training in radiochemistry. One of the most useful practical manuals for people in this category is *Radioisotope Laboratory Techniques*, [63] by Faires and Parks, which is based on experience at the UKAEA Isotope School, and gives guidance on techniques, equipment, procedures, hazard control and the disposal of waste. On similar lines is Smullen's *Basic Foundations of Isotope Technique for Technicians*. [64] At a more advanced level, *Radioisotope Techniques*, [65] by Overman and Clark, is a formal course for graduates, covering the handling of radioactive materials in the laboratory, measurement techniques and the interpretation of results. For those who need to delve more deeply into the radiochemical background, the practical aspects are discussed in *Modern Radiochemical Practice* [66] by Duncan and Cook, and in a more recent work [67] by Ladd and Lee, and complete coverage of every important concept and technique in this field is given in *Nuclear and Radiochemistry* [68] by Friedlander and others.

The design of radiochemical laboratories and the selection of equipment is the subject of *Hot Laboratory Equipment*. [69] edited by Stang, which includes information on protective clothing and shielding materials. This subject is well covered in a USAEC bibliography [70] published in 1960 and a comprehensive account of the problems of manipulating radioactive materials under shielded conditions is given in *Glove Boxes and Shielded Cells*, [71] edited by Walton.

LARGE RADIATION SOURCES

Radioisotopes, nuclear reactors and particle accelerators can be used as radiation sources where intense radiation is required as a means of modifying materials. Examples of such applications are sterilization of equipment, food preservation, plant breeding, insect control and chemical processing. All these are described in *Massive Radiation Techniques*, [72] a practical introduction which includes chapters on radiation sources and dosimetry. Industrial applications of this kind were the subject of two IAEA conferences, the proceedings [73] of which contain some papers on sources. The latter are discussed in more detail in a review [74] edited by Charlesby. One of the most promising aspects of this technique is in the field of food preservation and sterilization and this is the subject of a well-documented monograph [75] by Desrosier and Rosenstock.

EFFECTS OF RADIATION ON MATERIALS

Materials undergo physical and chemical changes when subjected to intense radiation and the study of these changes is important on two counts. First it is necessary to know how a material will behave in a radiation environment; for example, the reactor engineer must know how certain structural materials will stand up to the intense radiation flux in a nuclear reactor. Secondly, radiation can be used to modify physical or chemical structure and thus produce new materials which may be suitable for particular purposes.

The basic study in this field is that of radiation chemistry to which a comprehensive introduction has recently appeared. [76] At a more advanced level, radiation effects in gases are reviewed in *Radiation Chemistry of Gases* [77] which cites all significant experiments in the field during the period 1928 to 1960 and is amply provided with references to original papers. Swallow's *Radiation Chemistry of Organic Compounds* [78] reviews published work on this subject over the period 1895 to 1958, excluding reports and dissertations, and is fully documented. The effect of radiation on organic compounds is carried a step further in Charlesby's *Atomic Radiation and Polymers* [79] which examines the theoretical and experimental aspects od radiation-induced changes in organic molecules and particularly long-chain polymers. The most comprehensive bibliography [80] on radiation chemistry is that issued in six parts by AERE, Harwell which, with its supplements, covers the literature up to 1960.

Two books which are primarily of interest to engineers and experimentalists are *Radiation Effects on Organic Materials*, [81] by Bolt and Carroll, and *Effects of Radiation on Materials and Components,* [82] by Kircher and Bowman. The first of these deals with commonly used organic substances and gives enough information for the engineer to be able to predict the nature of radiation effects on these substances under given conditions. The second covers a wide range of materials including polymers, fuels, ceramics, metals and alloys and electronic components and is based on information accumulated at the Radiation Effects Information Centre at Battelle Memorial Institute.

BIOLOGICAL EFFECTS OF RADIATION

General

Radiobiology, the interaction between radiation and living organisms, is a subject which extends far beyond the boundaries of atomic energy. All radiation, from radio waves to cosmic rays, has some effect upon living cells if present in sufficiently large doses; here only ionizing radiation, i.e. alpha-, beta-, and gamma-rays, X-rays, protons and neutrons, will be considered and then only in so far as the phenomena touched upon have some reference to the general field of atomic energy.

The major effect of relatively large doses of ionizing radiation on a living organism is the destruction of cells which may lead to injury or death of the organism. If the irradiated cells happen to be reproductive cells even very small doses of radiation can cause changes in the genes (mutations) or in the arrangement of genes in the chromosomes, although the ability of the cells to live and divide is not impaired. In both cases the effects will be passed on to the daughter cells and the genetic pattern of the organism will be changed.

Radiation damage may therefore be studied at the molecular and cellular level, at the organism level and, in the case of long-term studies, at the genetic level and this division is to some extent reflected in the literature. Lea's *Actions of Radiations on Living Cells* [83] deals with phenomena at the molecular and cellular level and has become a standard introduction to the subject, suitable for both physicist and biologist. The book reviews earlier work in the field and has a long bibliography of over twenty pages. More recent information on phenomena at this level will be found in the twenty-four papers, all by acknowledged experts, contributed to a symposium [84] held in Moscow in 1960.

There are two comprehensive surveys of the whole field of radiobiology covering phenomena at all levels. *Radiation Biology*, [85] edited by Hollaender, is a three-volume work dealing with the effects, not only of ionizing radiation, but of ultra-violet and visible light and other forms of electromagnetic radiation. *Fundamentals of Radiobiology,* [86] by Bacq and Alexander, is a standard work on the subject which surveys the most important aspects and includes information on the pathology of radiation sickness in mammals and experimental work in the development of chemical protection methods. The book is fully documented and is intended mainly for physicians and radiotherapists.

A useful reference book on this subject is the *Atomic Energy Encyclopaedia in the Life Sciences* [87] which also includes materials on radiological protection, environmental contamination, waste disposal and the use of radiation and isotopes in medicine, agriculture and biology. Emphasis is mainly on American activities and developments and there are few references to original publications. A rather similar publication, although not in encyclopaedic form, is *Radiation Biology and Medicine* [88] which covers the same subject field in a series of carefully prepared reviews and is fully documented. For those who need more detailed information on radiation effects on particular organisms the USAEC have issued an annotated bibliography [89] containing over 11,000 references to reports and articles grouped on the basis of the type of organism affected.

Effects of Radiation on Man

The most important aspect of this subject is, of course, the effects of radiation on man, knowledge of which has been derived mainly from radiotherapy, from occupational exposure of workers and from the two atomic bomb explosions in Japan, particularly the latter. A Joint Commission for the Investigation of the Effects of the Atomic Bombs in Japan was set up shortly after the events occurred and its findings were published in one of the volumes [90] in the *National Nuclear Energy Series*. The report gives a detailed account of the damage caused and the general medical effects with a full description of the clinical observations made and the hematology and pathology of atomic bomb injuries.

In 1958 the United Nations published its first report [91] on the effects of atomic radiation which contains data on the observed levels of ionizing radiation in the human environment and describes observations and experiments relating to the effects of this radiation on man.

A second report, [92] issued in 1962, includes a much greater volume of data, collected from many countries, on all aspects of the problem. This report is a useful source of factual information on environmental radioactivity and the effects of radiation and is well documented.

Largely as a result of fears expressed about the danger from fall-out from nuclear explosions a number of countries set up machinery to examine the effects of radiation from natural and artificial sources and to keep these under continual review. In Britain the Medical Research Council published its first report in 1956, although the data available at this time was somewhat meagre. A second report [93] issued in 1960, based on a very much larger body of information, discusses the pathological and genetic effects of radiation and estimates the existing and foreseeable levels of exposure to radiation from all sources. One of the most useful features of the report is the appendices, each of which deal in detail with a specific aspect of the subject and all of which are fully documented. In America similar reports were issued by the National Research Council including a general report to the public published in 1956. [94]

Apart from its technical aspects, fall-out from nuclear explosions is, of course, a social and political problem and has received a good deal of attention in the literature. Over 5000 references are recorded in a USAEC bibliography [95] on the subject which covers the period up to February 1962 and includes both reports and journal articles. Both technical and political aspects are discussed in *Fall-out*, [96] which is a collection of essays on the nature of fall-out and the effects of radiation on human beings, and is basically a plea for the cessation of weapon testing. Fall-out and the initial effects of radiation from nuclear weapons are described in Glasstone's *Effects of Nuclear Weapons* [97] which also includes information on blast and thermal effects on structures

The radioactive fission products produced by a nuclear explosion or reactor excursion, and distributed over a wide area as a result of fall-out, are taken up by plants through the leaves or from the soil and thence enter the human food chain. The main contaminants are iodine-131, caesium-137, strontium-89 and strontium-90 and the metabolic behaviour of these isotopes in plants and animals is described in a report [98] issued by the Food and Agriculture Organization. The report suggests remedial measures and makes recommendations for combating this hazard but its chief value is in the factual information contained in the appendices, each of which includes a bibliography. Environmental radioactivity is discussed in more detail in Eisenbud's book, [99] which contains a useful section on protection standards and regulatory practices in the U.S.A. and a very long bibliography, and in the four papers [100] presented at a Royal Society of Health symposium in 1963, which describe the methods used for measuring and recording radioactive contamination of air, water and food.

The medical aspects of radiation damage are discussed in a number of textbooks and monographs. *Atomic Medicine*, [101] edited by Behrens, outlines the main effects of radiation and discusses the treatment of radiation sickness and injury and the protective measures which can be taken. The book also has a section on the applications of radioisotopes and radiation sources in medicine and biology. Browning's *Harmful Effects of Ionizing Radiations* [102] is a handbook for medical officers and safety officers in industry and gives a broad treatment with adequate references to important publications. *Radiation Injury in Man*, [103] a semi-technical presentation for those without special training in radiobiology, has a useful bibliography and is within the understanding of the general reader.

Popular Books

Radiobiology is another field which has captured the interest of the general public and there are a number of books suitable for the layman. Spear's *Radiations and Living Cells* [104] is a short introduction which emphasizes the effects on man, whereas *Atomic Radiation and Life*, [105] by Alexander, covers the whole field from radiation damage in individual cells to sickness in the whole animal and is extremely readable. *Radiation* [106] by Schubert and Lapp, examines not only the radiation hazards from fall-out and occupational exposure but those from the misuse of radiation in medical diagnosis and therapy and in other everyday contexts, and has a long bibliography of the important publications on the subject. Loutit, in his *Irradiation of Mice and Men* [107] makes a critical examination of what is known about the biological effects of radiation with particular reference to cancer, leukemia and longevity in man and devotes considerable space to the problems arising from the uptake of strontium-90 into the human food chains.

All the above contain sections on the genetic effects of radiation, particularly the book by Alexander. [105] A more detailed account is given in *Radiation, Genes and Man*, [108] by Wallace and Dobzhansky, and in *Genetic Effects of Radiation*, [109] by Purdom, which, although suitable for the intelligent layman, is also a useful introduction to the subject for students of biology.

PROTECTION AGAINST RADIATION

General

Although radiation damage, and particularly genetic damage, can, in theory, be caused by a single ionizing particle, in practice the probability of such damage being caused is proportional to the dose of radiation received and the main concern of the health physicist is to limit the dose which any worker may receive to a figure which ensures that this probability is small. The field of radiological protection then is concerned with defining levels of radiation which are an acceptable risk, with devising protective devices and methods of operation which ensure that such levels are not exceeded, and with designing and using measuring instruments which confirm that satisfactory conditions have been achieved.

All these aspects receive attention in a broad introduction to the subject by Eaves [110] which includes chapters on remote handling of radioactive material and waste disposal. A more detailed survey is made by Barnes and Taylor in *Radiation Hazards and Protection* [111] which is both a comprehensive text for the health physicist in training and a reference manual designed to help solve the day-to-day problems of those who do not have a trained person to advise them. *Protection against Radiation* [112] by Abbatt and others, is a practical handbook for those, with some biology but little physics, who may be responsible for the welfare of radiation workers. *Radiation Protection* [113] by Braestrup and Wyckoff contains a comprehensive summary of the effects of radiation, discusses shielding methods and materials and the principles of radiation protection in the various fields in which radiation is used with the exception of nuclear reactors. Shielding and protection in the environment of a reactor is discussed in the publications mentioned under the heading *Reactor Safety* in Chapter 9.

A useful reference book in this field is the *Radiation Hygiene Handbook* [114] which covers the industrial, medical and research uses of radiation with special emphasis on health physics, industrial hygiene and sanitation. *Radiation and Health*, [115] edited by Williams and others, also stresses the public health aspects of the subject and is based on papers presented at a meeting of Medical Officers of Health, held in London in 1958. The book is particularly concerned with protection against environmental radiation arising from nuclear explosions, large scale use of radioactive substances and other sources.

A number of handbooks are available to give guidance on the protective measures necessary in handling radioactive substances on a large or small scale. The Medical Research Council's *Introductory Manual* [116] is a useful handbook for anyone handling radioisotopes in the laboratory and is relevant to industrial, medical and biological applications. Quimby's *Safe Handling of Radioactive Isotopes* [117] is designed for doctors, nurses, technicians and other hospital staff, and the *Manual of Radioactivity Procedures* [118] issued by the National Bureau of Standards is also of interest mainly to those in the medical field and is particularly useful for those concerned with the disposal of very low activity wastes.

Apart from physical methods of protection a good deal of experimental work has been done on chemical and biochemical methods, i.e. the use of chemicals before exposure to radiation in order to reduce the effects of the dose. This subject is discussed authoritatively in the twelve papers which make-up *Radiation Protection and Recovery*, [119] all of which are fully documented.

Codes of Practice

The important principles of radiological protection have been embodied in a number of codes of practice, at both national and international level, which seek to:

(1) Define the maximum permissible dose which a person may receive without his health being affected.
(2) Make recommendations for the practical control of radiation dose by setting out the operating conditions which must be observed and tabulating the maximum permissible concentrations of various radionuclides in the environment.

The main set of recommendations, from which nearly all the others are derived, is that of the International Commission on Radiological Protection. [120] The recommendations give maximum permissible doses for whole-body irradiation and for separate parts of the body and suggest a model code of practice relating to working conditions which other organizations can follow.

The European Nuclear Energy Agency has published its own set of standards, [121] which it expects member states to adopt, and the International Atomic Energy Agency has issued *Basic Safety Standards* [122] which are applied in Agency-sponsored projects and which, it is hoped, will serve as a basis for the formulation of national regulations in individual countries. The IAEA has also published a code of practice for handling isotopes, [123] based on the ICRP recommendations and designed mainly for small users, which includes two addenda giving some account of the technical background, an understanding of which is necessary for proper implementation of the code. In the industrial field the International Labour Office publishes a *Manual* [124] which includes a model code of safety regulations and a general guide to radiological protection.

Many countries have enacted legislation dealing with radiation hazards and published regulations and codes of practice for use at the national level. British legislation is discussed in the books already mentioned by Eaves [110] and Williams [115] and a summary of existing legislation and possible future enactments in Britain appeared in an article [125] in *Annals of Occupational Hygiene* in 1963 which was subsequently reprinted in *Atom*. Regulations relating primarily to the use of radioactive materials in teaching establishments are reviewed by Roberts in an article in *Contemporary Physics* [126] and a code of practice [127] for persons working in industrial research laboratories, teaching laboratories and research establishments (other than hospitals) has been issued by the British Government. A list [128] of British official and semi-official publications, including acts, regulations and codes of practice, has been issued by the Radiological Protection Division of the UKAEA Health and Safety Branch.

A summary of American legislation and regulations appears in Eisenbud's *Environmental Radioactivity*, [99] which has already been mentioned. In America the main body is the National Committee on Radiation Protection and Measurement whose recommendations are issued periodically by the National Bureau of Standards. The USAEC's regulatory programme is based on publications issued as part of the Code of Federal Regulations. [129] There are, however, a large number of American organizations with responsibilities in this field, most of which are listed in the *Selected Bibliography of Radiation Protection Organizations*, [130] which gives details of their responsibilities, activities and publications. Many of the latter are included in a USAEC bibliography [131] of reports and journal articles on radiation protection standards.

The problems associated with the transport and handling of active materials are reviewed by Farmer [132] in an IAEA publication issued in 1961 which considers not only the problem of radiation protection but also that of transporting sub-critical quantities of fissile materials with safety. The IAEA have also issued *Regulations for the Safe Transport of Radioactive Materials* together with notes [133] which give guidance on the purpose of the regulations and their technical background and include a layman's guide in the form of question and answer.

There is in fact a considerable literature on this subject and reference should be made to a USAEC bibliography [134] issued in 1961 which lists mainly American papers and reports published between 1950 and 1960.

Measurement of Radiation Dose

All protective measures rely on the ability to measure the quantity of radiation received or present in particular circumstances. *Radiation Dosimetry*, [135] by Hine and Brownell, is a unified presentation of the fundamental principles of dosimetry and a description of the measuring instruments available, with full references to earlier work on the subject. The basic principles are set out coherently in Whyte's book, [136] designed for the scientist faced with the problem of determining radiation dose in given circumstances, whereas Handloser's *Health Physics Instrumentation* [137] describes the counters, detectors, monitors and associated equipment of interest to the health physicist. Basically these instruments are adaptations of the counters and detectors which are in everyday use in nuclear physics and atomic energy research and which form the subject of the next section.

RADIATION COUNTERS AND DETECTORS

Since ionizing radiations are completely undetectable by human senses, instruments are an essential feature of all work in the atomic energy field, particularly in experimental nuclear physics and in studies involving radioactive materials.

Nuclear Radiation Detection, [138] by Price, is a collection of basic information on all types of radiation detector. With copious references to earlier publications, it covers conventional devices based on the use of ionization chambers such as Geiger and proportional counters, solid state devices such as scintillation counters, and photographic emulsions and it pays some attention to the statistical aspects of particle counting. Sharpe's *Nuclear Radiation Detectors* [139] covers much the same ground but emphasizes the physical principles on which the operation of detector elements is based and is of interest mainly to the designer of counting instruments.

A simple and practical introduction to the use and operation of the more conventional counters is provided by Washtell [140] and more detailed information on their principles of operation and limiting characteristics is given in Wilkinson's *Ionization Chambers and Counter* [141] which has a long list of references.

Counters based on the use of an ionization chamber are not particularly efficient for detecting penetrating radiation such as X-rays and gamma-rays and solid or liquid detectors are used for this purpose. Of these, scintillation counters, which depend on the production of light by fast particles, are the subject of comprehensive monographs by Birks [142] and Curran, [143] both of which are well documented. In general scintillation counters are based on solid luminescent materials used in conjunction with a photomultiplier to register the light flashes but in some circumstances increased sensitivity may be obtained by the use of liquid scintillators. The problems associated with the use of liquids are discussed in the proceedings [144] of a conference held at Northwestern University in 1957.

A similar type of detector is the Cerenkov counter which depends on the fact that particles travelling in a medium at a velocity greater than that of light in the medium emit a luminous radiation which can be used to excite a photomultiplier. The general properties and main applications of Cerenkov detectors are described by Jelley in *Cerenkov Radiation*, [145] the only monograph available on this subject.

More recently a new type of solid state detector is beginning to emerge based on the use of semiconductors. Taylor's *Semiconductor Particle Detectors* [146] covers both theoretical and practical aspects with emphasis on the general properties of semiconductors and their present limitations. A long bibliography of earlier work in this field will be found in *Semiconductor Counters for Nuclear Radiations* [147] by Dearnaley and Northrop. A number of original papers on developments in solid state detectors is contained in the proceedings [148] of a symposium on nuclear instruments held at Harwell in 1961.

Neutrons cannot be detected directly in instruments designed for counting charged particles so that special detectors are needed for neutron measurement. *Neutron Detection*, [149] by Allen, is based on information available up to 1957 and is intended for those, with some background in nuclear physics, who need a detailed knowledge of neutron detection methods. A very good bibliography and copious references enable it to be used by the more expert experimenter as a handy reference book.

The electrical signals generated in a detector are usually weak and require amplification for use in any practical situation. The design of particle detector amplifiers is discussed by Gillespie [150] in a classic work which pays particular attention to the problems of amplifier noise and signal-to-noise ratio. Of course, where situations arise in which it is necessary to measure weak levels of radioactivity, close to the natural background, special techniques are needed at the pre-amplification stage and these are described by Watt and Ramsden in *High Sensitivity Counting Techniques*. [151]

In addition to the general works on counters so far mentioned there are a number of books written for special classes of users. The first volume of *Nuclear Instruments and their Uses*, [152] edited by Snell, is mainly concerned with dosimetry and health physics, although all the main types of counter are mentioned. Taylor's *Measurement of Radioisotopes* [153] is designed for the non-specialist working with radioactive materials and the author has taken great care to explain the advantages and limitations of each method described. Fremlin's *Applications of Nuclear Physics* [154] discusses instruments and measuring techniques in the radioisotopes field and is also intended for non-specialists.

Cloud chambers, bubble chambers and other instruments and techniques used in nuclear and high energy physics are described by Ritson [155] in a book which also contains information on the use of photographic emulsions for the detection of individual particles. This is a technique used mainly in elementary particle physics and cosmic ray studies, as described by Powell and others, [156] but it can also be used as a quantitative counting method for measuring radioisotopes, particularly those which decay by alpha-emission, and this aspect has been dealt with fairly comprehensively by Yagoda. [157]

REFERENCES

General

1. Lapp, R. E. and Andrews, H. L. *Nuclear radiation physics.* 3rd ed. London, Pitman, 1964.
2. Whitehouse, W. J. and Putman, J. L. *Radioactive isotopes: an introduction to their preparation, measurement and use.* Oxford, Clarendon Press, 1953.
3. Rotblat, J. and others. *The uses and effects of nuclear energy.* London, Harrap, 1964.
4. Putman, J. L. *Isotopes.* Harmondsworth, Penguin Books, 1960.
5. Glascock, R. *Labelled atoms: the use of radioactive and stable isotopes in biology and medicine.* London, Sigma Books, 1951.
6. Fearnside, K., Jones, E. W. and Shaw, E. N. *Applied atomic energy.* London, Temple Press, 1951.
7. Jefferson, S. *Radioisotopes: a new tool for industry.* 2nd ed. London, Newnes, 1960.

Isotope Separation

8. Aston, F. W. *Mass spectra and isotopes.* 2nd ed. London, E. Arnold, 1942.
9. Smith, M. L., ed. *Electromagnetically enriched isotopes and mass spectrometry: proceedings of the conference held at Harwell, September, 1955.* London, Butterworth, 1956.
10. Higatsberger, M. J. and Viehböck, F. P., eds. *Electromagnetic separation of radioactive isotopes: proceedings of the International Symposium held in Vienna, May, 1960.* Vienna, Springer-Verlag, 1961. (In English.)
11. Koch, J. and others. *Electromagnetic isotope separators and applications of electromagnetically enriched isotopes.* Amsterdam, North Holland, 1958.
12. Kistemaker, J. and others, eds. *Proceedings of the International Symposium on Isotope Separation, held in Amsterdam, April, 1957.* Amsterdam, North Holland, 1958.
13. London, H., ed. *Separation of isotopes.* London, Newnes, 1961.
14. International Atomic Energy Agency. *Production and use of shortlived radioisotopes from reactors: proceedings of the seminar ... held by the IAEA, Vienna, 1962.* Vienna, IAEA, 1963.

Applications of Isotopes and Radiations

15. United States Atomic Energy Commission, Division of Isotopes Development. *Special sources of information on isotopes in industry, agriculture, medicine and research.* USAEC, Division of Technical Information Extension, P.O. Box 62, Oak Ridge, Tennessee, April, 1962. (TID—4563, 3rd rev.)
16. United States Atomic Energy Commission, Division of Isotopes Development. *Radioisotopes in world industry: abstracts of selected foreign literature.* Washington, Office of Technical Services, January 1961. (TID—6613.) Supplements issued in June 1961, January 1962, October 1962, and July 1963.
17. United Kingdom Atomic Energy Authority, Radiochemical Centre. *Selected references to tracer techniques.* Amersham (Bucks), Radiochemical Centre, July 1964. (RCC Review 1.)
18. Ministry of Supply. *Radioisotope techniques: proceedings of the Isotope Techniques Conference, Oxford, July, 1951.* London, H.M.S.O., 1953, 2 vols.
    Vol. 1: Therapy, diagnosis, biochemistry, agriculture; Vol. 2: Industrial and allied research.
19. Johnston, J. E., Faires, R. A. and Millet, R. J. *Radioisotope conference 1954: proceedings of the second conference, Oxford, July, 1954.* London, Butterworth, 1954. 2 vols.

Vol. 1: Medical and physiological aspects; Vol. 2: Physical sciences and industrial applications.
20. EXTERMANN, R. C., ED. *Radioisotopes in scientific research: proceedings of the international conference held in Paris, September, 1957.* London, Pergamon Press, 1958. 4 vols.
Vol. 1: Physics and industry; Vol. 2: Chemistry and geology; Vol. 3: Human and animal biology; Vol. 4: Plant biology and general problems.
21. INTERNATIONAL ATOMIC ENERGY AGENCY. *Radioisotopes in the physical sciences and industry: proceedings of the conference ... held by the IAEA with the cooperation of UNESCO at Copenhagen, September, 1960.* Vienna, IAEA, 1962. 3 vols.

Applications in Physical Sciences and Industry

22. ILLINOIS INSTITUTE OF TECHNOLOGY, RESEARCH INSTITUTE. *Literature survey on world isotope and radiation technology: final report June, 1961 to June, 1962.* Washington, Office of Technical Services, 1963. (IITRI—1194—13.)
23. STUART, D. F. O. AND BIRCH, F. Survey on the use of radioisotopes in British industry and industrial research. *Atom,* no. 94, pp. 176—190, August 1964.
24. UNITED STATES ATOMIC ENERGY COMMISSION. *Radioisotopes in science and industry: a special report of the USAEC.* Washington, U.S. Government Printing Office, 1960.
25. INTERNATIONAL ATOMIC ENERGY AGENCY. *Radioisotope applications in industry.* Vienna, IAEA, 1963.
26. ROCHLIN, R. S. AND SCHULTZ, W. W. *Radioisotopes for industry.* New York, Reinhold; London, Chapman and Hall, 1959.
27. KOHL, J., ZENTNER, R. D. AND LUKERS, H. R. *Radioisotope applications engineering.* Princeton, Van Nostrand, 1961.
28. SHUMILOVSKII, N. N. AND MEL'TTSEN, L. V. *Radioactive isotopes in instrumentation and control,* translated by R. F. Kelleher. Oxford, Pergamon Press, 1964.
29. HART, H. AND OTHERS. *Radioaktive Isotope in der Betriebsmeßtechnik (Radioactive isotopes in industrial measurement techniques).* Berlin, VEB Verlag Technik, 1962. (In German.)
30. (a) CAMERON, J. F. Backscattering gauges. *Atom,* no. 55, pp. 11—12, bibliog., May 1961.
(b) CAMERON, J. F. Transmission gauges. *Atom,* no. 55, pp. 18—19, bibliog., May 1961.
31. ROCKLEY, J. C. *An introduction to industrial radiology.* London, Butterworth, 1964.
32. DAUDEL, P. *Radioactive tracers in chemistry and industry.* London, Griffin, 1960.
33. UNITED KINGDOM ATOMIC ENERGY AUTHORITY, RADIOCHEMICAL CENTRE. *Radioactive isotope dilution analysis.* Amersham (Bucks), Radiochemical Centre, April 1964. (RCC Review 2.)
34. LAMBIE, D. A. *Techniques for the use of radioisotopes in analysis: a laboratory manual.* London, Spon, 1964.
35. TAYLOR, D. *Neutron irradiation and activation analysis.* London, Newnes, 1964.
36. BOWEN, H. J. M. AND GIBBONS, D. *Radioactivation analysis.* Oxford, Clarendon Press, 1963.

Biological Applications

37. KUZIN, A. M. *The application of radioisotopes in biology.* (In English and Russian.) Vienna, IAEA, 1960. (Review series no. 7.)
38. HEVESY, G. *Radioactive indicators; their application in biochemistry, animal physiology and pathology.* New York, Interscience, 1948.

39. COMAR, C. L. *Radioisotopes in biology and agriculture: principles and practice.* New York, McGraw-Hill, 1955.
40. KAMEN, M. D. *Isotopic tracers in biology: an introduction to tracer methodology.* 3rd ed. New York, Academic Press, 1957.
41. FRANCIS, G. E., MULLIGAN, W. AND WORMALL, A. *Isotopic tracers: a theoretical and practical manual for biological students and research workers.* 2nd ed. University of London, Athlone Press, 1959.
42. SACKS, J. *Isotopic tracers in biochemistry and physiology.* New York, McGraw-Hill, 1953.
43. BRODA, E. *Radioactive isotopes in biochemistry.* Amsterdam, London, New York; Elsevier, 1960.
44. *Radioactive isotopes in medicine and biology.* 2nd ed.
Vol. 1: Basic physics and instrumentation, by E. H. Quimby and S. Feitelberg. London, Kimpton, 1963; Vol. 2: Medicine, by S. Silver, London, Kimpton, 1962.
45. VEALL, N. AND VETTER, H. *Radioisotope techniques in clinical research and diagnosis.* London, Butterworth, 1958.
46. BEIERWALTES, W. H. AND OTHERS. *Clinical use of radioisotopes.* Philadelphia, London, W. B. Saunders, 1957.
47. HALNAN, K. E. *Atomic energy in medicine.* London, Butterworth, 1957.
48. KING, E. R. AND MITCHELL, T. G. *A manual for nuclear medicine.* Springfield (Illinois), C. C. Thomas, 1961.
49. FIELDS, T. AND SEED, L., EDS. *Clinical use of radioisotopes: a manual of technique.* 2nd ed. Chicago, Year Book Publishers, 1961.
50. DELARIO, A. J. *Roentgen, radium and radioisotope therapy.* London, Kimpton, 1953.
51. PATERSON, R. *The treatment of malignant disease by radiotherapy.* 2nd ed. London, E. Arnold, 1963.
52. UNITED STATES ATOMIC ENERGY COMMISSION, DIVISION OF TECHNICAL INFORMATION. *Radioisotopes in medicine and human physiology: a selected list of references.* Washington, Office of Technical Services, 1948—1960.
Part 1: TID—3514, August 1958; Part 2: TID—3077, March 1960.
53. ETTER, L. E. *Glossary of words and phrases used in radiology and nuclear medicine.* Springfield (Illinois), C. C. Thomas, 1960.
54. DICK, W. E. *Atomic energy in agriculture.* London, Butterworth, 1957.
55. SINGLETON, W. R. *Nuclear radiation in food and agriculture.* Princeton, Van Nostrand, 1958.
56. KAINDL, K. AND LINSER, H. *Radiation in agricultural research and practice.* Vienna, IAEA, 1961. (Review series, no. 10.)
57. (a) UNITED STATES ATOMIC ENERGY COMMISSION, DIVISION OF TECHNICAL INFORMATION. *Radioisotopes in agriculture: animal husbandry, bacteriology, fertilizer uptake, plant physiology, photosynthesis and entomology,* compiled by J. A. M. McCormick. Washington, Office of Technical Services, 1959. (TID—3516.)

(b) UNITED STATES ATOMIC ENERGY COMMISSION, DIVISION OF TECHNICAL INFORMATION. *Radioisotopes in agriculture: animal husbandry, fertilizer uptake, plant physiology, photosynthesis and entomology,* a selected bibliography compiled by J. A. McCormick and H. E. Voress. Washington, Office of Technical Services, 1960. (TID—3078.)

(c) UNITED STATES ATOMIC ENERGY COMMISSION, DIVISION OF TECHNICAL INFORMATION. *Radioisotopes in agriculture: analytical procedures, animal husbandry, entomology, fertilizer uptake, general studies, photosynthesis, plant genetics and plant physiology,* a selected bibliography compiled by H. L. Wort. Washington, Office of Technical Services, 1964. (TID—3078, suppl. 1.)

Directories and Catalogues of Isotopes

58. INTERNATIONAL ATOMIC ENERGY AGENCY. *International directory of radioisotopes.* 2nd ed. Vienna, IAEA, 1962.
Part 1: Unprocessed and processed radioisotope preparations and special radiation sources; Part 2: Compounds labelled with C^{14}, H^3, I^{131}, P^{32} and S^{35}.
59. UNITED KINGDOM ATOMIC ENERGY AUTHORITY, RADIOCHEMICAL CENTRE. *Catalogue of radioactive products.* Amersham (Bucks), Radiochemical Centre, 1964.
60. UNITED KINGDOM ATOMIC ENERGY AUTHORITY, RADIOCHEMICAL CENTRE. *The radiochemical manual.* Part 1, Physical data. Part 2, Radioactive chemicals. London, H.M.S.O., 1962. 2 vols.
61. SCIENTIFIC EQUIPMENT CO. *Isotope index 1963—1964.* Scientific Equipment Co., Publications Dept., P.O. Box 19086, Indianapolis 19, Indiana, 1963.
62. (a) OAK RIDGE NATIONAL LABORATORY. *Radioisotopes, special materials and services: catalog and price list.* 3rd rev. Oak Ridge National Laboratory, Tennessee, 1960.
(b) OAK RIDGE NATIONAL LABORATORY. *Catalog and price list of stable isotopes, including selected materials and services.* Oak Ridge National Laboratory, Tennessee, 1961.

Isotope Techniques

63. FAIRES, R. A. AND PARKS, B. H. *Radioisotope laboratory techniques.* 2nd ed. London, Newnes, 1960.
64. SMULLEN, W. C., ED. *Basic foundations of isotope technique for technicians.* Springfield (Illinois), C. C. Thomas, 1956.
65. OVERMAN, R. T. AND CLARK, H. M. *Radioisotope techniques.* New York, McGraw-Hill, 1960.
66. DUNCAN, J. F. AND COOK, G. B. *Modern radiochemical practice.* Oxford, Clarendon Press, 1952.
67. LADD, M. F. C. AND LEE, W. H. *Practical radiochemistry.* London, Cleaver-Hume, 1964.
68. FRIEDLANDER, G., KENNEDY, J. W. AND MILLER, J. M. *Nuclear and radiochemistry.* 2nd ed. New York, Wiley, 1964.
69. STANG, L. G. *Hot laboratory equipment.* Washington, U.S. Government Printing Office, 1958.
70. UNITED STATES ATOMIC ENERGY COMMISSION, DIVISION OF TECHNICAL INFORMATION. *Hot laboratories: a literature search,* compiled by R. L. Scott. Washington, Office of Technical Services, 1960.
71. WALTON, G. N., ED. *Glove boxes and shielded cells for handling radioactive materials.* New York, Academic Press, 1958.

Large Radiation Sources

72. JEFFERSON, S., ED. *Massive radiation techniques.* London, Newnes, 1964.
73. (a) INTERNATIONAL ATOMIC ENERGY AGENCY. *Large radiation sources in industry: proceedings of a conference on the application of large radiation sources in industry and especially to chemical processes, organised by the IAEA, Warsaw, 1959.* Vienna, IAEA, 1960. 2 vols.
(b) INTERNATIONAL ATOMIC ENERGY AGENCY. *Industrial uses of large radiation sources: proceedings of a conference on the application of large radiation sources in industry, held by the IAEA, Salzburg, 1963.* Vienna, IAEA, 1963. 2 vols.
74. CHARLESBY, A., ED. *Radiation sources.* Oxford, Pergamon Press, 1964.
75. DESROSIER, N. W. AND ROSENSTOCK, H. M. *Radiation technology in food agriculture and biology.* Westport (Conn.), AVI Publ. Co., 1960.

Effects of Radiation on Materials

76. SPINKS, J. W. T. AND WOODS, R. J. *An introduction to radiation chemistry.* New York, Wiley, 1964.
77. LIND, S. C. AND OTHERS. *Radiation chemistry of gases.* New York, Reinhold; London, Chapman and Hall, 1961.
78. SWALLOW, A. J. *Radiation chemistry of organic compounds.* Oxford, Pergamon Press, 1960.
79. CHARLESBY, A. *Atomic radiation and polymers.* Oxford, Pergamon Press, 1960.
80. ATOMIC ENERGY RESEARCH ESTABLISHMENT. *Selected abstracts of atomic energy project unclassified report literature in the field of radiation chemistry and bibliography of the published literature,* by R. W. Clarke. London, H.M.S.O., 1956. (AERE C/R 1575.)
    Part 1: Theory, interpretations, water and aqueous inorganic systems; Part 2: Organic compounds, including polymerization reactions; Part 3: Gaseous systems; Part 4: Solid systems; Part 5: Biochemistry and radiobiology; Part 6: Miscellaneous.
    Annual supplements issued in 1957, 1958, 1959, 1960 and 1961.
81. BOLT, R. O. AND CARROLL, J. G. *Radiation effects on organic materials.* New York, Academic Press, 1963.
82. KIRCHER, J. F. AND BOWMAN, R. E. *Effects of radiation on materials and components.* New York, Reinhold; London, Chapman and Hall, 1964.

Biological Effects of Radiation

83. LEA, D. E. *Actions of radiations on living cells.* 2nd ed. Cambridge University Press, 1955.
84. HARRIS, R. J. C., ED. *The initial effects of ionizing radiation on cells; proceedings of a symposium held in Moscow, 1960.* London, New York, Academic Press, 1961.
85. HOLLAENDER, A., ED. *Radiation biology.* New York, McGraw-Hill, 1954—1956. 3 vols.
86. BACQ, Z. M. AND ALEXANDER, P. *Fundamentals of radiobiology.* 2nd ed. Oxford, Pergamon Press, 1961.
87. SHILLING, C. W., ED. *Atomic energy encyclopaedia in the life sciences; prepared under the auspices of the Division of Technical Information, USAEC.* Philadelphia, London, W. B. Saunders, 1964.
88. CLAUS, W. D., ED. *Radiation biology and medicine; selected reviews in the life sciences.* Reading (Mass.), Addison-Wesley, 1958. (Atoms for Peace series.)
89. UNITED STATES ATOMIC ENERGY COMMISSION, DIVISION OF TECHNICAL INFORMATION. *The effects of radiation and radioisotopes on the life processes: an annotated bibliography.* Washington, Office of Technical Services, 1963. 3 vols.
90. OUGHTERSON, A. W. AND WARREN, S. *Medical effects of the atomic bomb in Japan.* New York, McGrave-Hill, 1956. (NNES series.)
91. UNITED NATIONS. *Report of the United Nations Scientific Committee on the Effects of Atomic Radiation.* New York, United Nations, 1958.
92. UNITED NATIONS. *Report of the United Nations Scientific Committee on the Effects of Atomic Radiation.* New York, United Nations, 1962. 2 vols.
93. MEDICAL RESEARCH COUNCIL. *The hazards to man of nuclear and allied radiations: a second report to the Medical Research Council.* London, H.M.S.O., 1960. (Cmnd. 1225.)
94. NATIONAL ACADEMY OF SCIENCES, NATIONAL RESEARCH COUNCIL. *The biological effects of atomic radiation: a report to the public.* Washington, National Research Council, 1956.

95. UNITED STATES ATOMIC ENERGY COMMISSION, DIVISION OF TECHNICAL INFORMATION. *Radioactive fall-out: a bibliography of the world's literature.* Washington, Office of Technical Services, 1961. (TID—3086.) Supplement 1 published July 1962. (TID—3086, suppl. 1.)
96. PIRIE, A., ED. *Fall-out: radiation hazards from nuclear explosions.* Rev. ed. London, MacGibbon and Kee, 1958.
97. GLASSTONE, S., ED. *The effects of nuclear weapons.* Rev. ed. U.S. Government Printing Office, Washington, 1962.
98. FOOD AND AGRICULTURE ORGANIZATION OF THE UNITED NATIONS. *Radioactive materials in food and agriculture; report of an FAO expert committee, Rome, 1959.* Rome, Food and Agriculture Organization, 1960. (FAO Atomic Energy Series, no. 2.)
99. EISENBUD, M. *Environmental radioactivity.* New York, McGraw-Hill, 1963.
100. MCLEAN, A. S. AND OTHERS. *Radiation levels: air, water and food.* London, Royal Society of Health, 1964.
101. BEHRENS, C. F., ED. *Atomic medicine.* 3rd ed. London, Bailliere, Tindall and Cox, 1959.
102. BROWNING, E. *Harmful effects of ionizing radiations.* Amsterdam, London, Elsevier, 1959.
103. CRONKITE, E. P. AND BOND, V. P. *Radiation injury in man: its chemical and biological basis, pathogenesis and therapy.* Springfield (Illinois), C. C. Thomas, 1960.
104. SPEAR, F. G. *Radiations and living cells.* London, Chapman and Hall, 1953.
105. ALEXANDER, P. *Atomic radiation and life.* Harmondsworth, Penguin Books, 1957.
106. SCHUBERT, J. AND LAPP, R. E. *Radiation: what it is and how it effects you.* New York, Viking Press, 1957.
107. LOUTIT, J. F. *Irradiation of mice and men.* University of Chicago Press, 1962.
108. WALLACE, B. AND DOBZHANSKY, T. *Radiation, genes and men.* New York, Henry Holt and Co., 1959.
109. PURDOM, C. E. *Genetic effects of radiations.* London, Newnes, 1963.

Radiological Protection

110. EAVES, G. *Principles of radiation protection.* London, Iliffe, 1964.
111. BARNES, D. E. AND TAYLOR, D. *Radiation hazards and protection.* 2nd ed. London, Newnes, 1963.
112. ABBATT, J. D., LAKEY, J. R. A. AND MATHIAS, D. J. *Protection against radiation.* London, Cassell, 1961.
113. BRAESTRUP, C. B. AND WYCKOFF, H. O. *Radiation protection.* Springfield (Illinois), C. C. Thomas, 1958.
114. BLATZ, H., ED. *Radiation hygiene handbook.* New York, McGraw-Hill, 1959.
115. WILLIAMS, K., SMITH, C. L. AND CHALKE, H. D., EDS. *Radiation and health.* London, Longmans, 1962.
116. MEDICAL RESEARCH COUNCIL, *Committee on Protection against Ionizing Radiations. Introductory manual on the control of health hazards from radioactive materials.* London, H.M.S.O., 1961. (Memo. no. 39.)
117. QUIMBY, E. N. *Safe handling of radioactive isotopes in medical practice.* New York, Macmillan Co., 1960.
118. NATIONAL BUREAU OF STANDARDS. *A manual of radioactivity procedures.* Washington, U.S. Government Printing Office, 1961. (NBS Handbook no. 80.)
119. HOLLAENDER, A., ED. *Radiation protection and recovery.* Oxford, Pergamon Press, 1960.

Codes of Practice and Regulations

120. INTERNATIONAL COMMISSION ON RADIOLOGICAL PROTECTION. *Radiation protection: recommendations of the ICRP (as amended 1959 and revised 1962).* Oxford, Pergamon Press, 1964. (ICRP pub. no. 6.)
121. EUROPEAN NUCLEAR ENERGY AGENCY. *Radiation protection norms.* Paris, OECD, 1963.
122. INTERNATIONAL ATOMIC ENERGY AGENCY. *Basic safety standards for radiation protection.* Vienna, IAEA, 1962. (Safety series no. 9.)
    Available in English, French, Russian and Spanish editions.
123. INTERNATIONAL ATOMIC ENERGY AGENCY. *Safe handling of radioisotopes.* First ed. with revised Appendix 1. Vienna, IAEA, 1962. (Safety series no. 1.) Health physics addendum. Vienna, IAEA, 1960. (Safety series no. 2.) Medical addendum. Vienna, IAEA, 1960. (Safety series no. 3.)
124. INTERNATIONAL LABOUR OFFICE. *Manual of industrial radiation protection.*
    Part 1. Convention and recommendations. 1963; Part 2. Model code of safety regulations. 1959; Part 3. General guide on protection against ionizing radiations. 1963; Part 4. Guide on protection against ionizing radiation in industrial radiography and fluoroscopy; Part 5. Guide on protection against ionizing radiations in the application of luminous compounds.
    Geneva, ILO, in progress.
125. SIM, D. F. The future of national and international regulations covering the use of ionizing radiations. *Annals of Occupational Hygiene,* vol. 7, pp. 63—69, January 1964. Reprinted in *Atom,* no. 89, pp. 55—60, March 1964.
126. ROBERTS, J. E. Codes of practice for the use of ionizing radiations. *Contemporary Physics,* vol. 4, no. 3, pp. 178—186, February 1963.
127. MINISTRY OF LABOUR. *Code of practice for the protection of persons exposed to ionizing radiations in research and teaching.* London, H.M.S.O., 1964.
128. WRAY, E. T., COMP. *Radiation protection: reference list of official and semi-official publications.* Radiological Protection Division, UKAEA Health and Safety Branch, AERE, Harwell, August 1964.
129. UNITED STATES ATOMIC ENERGY COMMISSION. *Rules and regulations. (Title 10, Code of Federal Regulations, Chapter 1).* Washington, USAEC, 1962 (with amendments).
130. AMERICAN CONFERENCE OF GOVERNMENTAL AND INDUSTRIAL HYGIENISTS, COMMITTEE ON IONIZING RADIATION. *Selected bibliography of radiation protection organisations.* The Conference, 1014 Broadway, Cincinnati, Ohio, 1963.
131. BOST, W. E. COMP. *Radiation protection standards: a literature search.* Washington, Office of Technical Services, June 1961. (TID—3551, rev. 1.)
132. FARMER, F. R. *The packaging, transport and related handling of radioactive materials.* Vienna, IAEA, 1961. (Review series no. 12.)
133. (a) INTERNATIONAL ATOMIC ENERGY AGENCY. *Regulations for the safe transport of radioactive materials.* Vienna, IAEA, 1961. (Safety series no. 6.)
    (b) INTERNATIONAL ATOMIC ENERGY AGENCY. *Regulations for the safe transport of radioactive materials: notes on certain aspects of the regulations.* Vienna, IAEA, 1961. (Safety series no. 7.)
134. UNITED STATES ATOMIC ENERGY COMMISSION, OFFICE OF TECHNICAL INFORMATION. *Shipping, handling and storage of radioactive materials: a literature search.* Washington, Office of Technical Services, 1961. (TID—3552, rev.)

Dosimetry

135. HINE, G. J. AND BROWNELL, G. L., EDS. *Radiation dosimetry.* New York, Academic Press, 1956.

136. WHYTE, G. N. *Principles of radiation dosimetry.* New York, Wiley; London, Chapman and Hall, 1959.
137. HANDLOSER, J. S. *Health physics instrumentation.* London, Pergamon Press, 1959.

Counters and Detectors

138. PRICE, W. J. *Nuclear radiation detection.* New York, McGraw-Hill, 1958.
139. SHARPE, J. *Nuclear radiation detectors.* 2nd ed. London, Methuen; New York, Wiley, 1964.
140. WASHTELL, C. C. H. *An introduction to radiation counters and detectors.* London, Newnes, 1958.
141. WILKINSON, D. H. *Ionization chambers and counters.* Cambridge University Press, 1950.
142. BIRKS, J. B. *Scintillation counters.* London, Pergamon Press, 1953.
143. CURRAN, S. C. *Luminescence and the scintillation counter.* London, Butterworth, 1953.
144. BELL, C. G. AND HAYES, F. N., EDS. *Liquid scintillation counting: proceedings of a conference held at Northwestern University, August, 1957.* London, Pergamon Press, 1958.
145. JELLEY, J. V. *Cerenkov radiation.* London, Pergamon Press, 1959.
146. TAYLOR, J. M. *Semiconductor particle detectors.* London, Butterworth, 1963.
147. DEARNALEY, G. AND NORTHROP, D. C. *Semiconductor counters for nuclear radiations.* London, E. and F. Spon, 1963.
148. BIRKS, J. B., ED. *Proceedings of the symposium on Nuclear Instruments, Harwell, 1961.* London, Heywood and Co., 1962.
149. ALLEN, W. D. *Neutron detection.* London, Newnes, 1960.
150. GILLESPIE, A. B. *Signal, noise and resolution in nuclear counter amplifiers.* London, Pergamon Press, 1953.
151. WATT, D. E. AND RAMSDEN, D. *High sensitivity counting techniques.* Oxford, Pergamon Press, 1964.
152. SNELL, A. H., ED. *Nuclear instruments and their uses. Vol. 1. Ionization detectors, scintillators, Cerenkov counters, amplifiers: assay, dosimetry, health physics.* New York, London, Wiley, 1962.
153. TAYLOR, D. *The measurement of radioisotopes.* 2nd ed. London, Methuen, 1957.
154. FREMLIN, J. H. *Applications of nuclear physics.* London, E.U.P., 1964.
155. RITSON, D. M. *Techniques of high energy physics.* London, New York, Interscience, 1961.
156. POWELL, C. F., FOWLER, P. H. AND PARKINS, D. H. *The study of elementary particles by the photographic method.* London, Pergamon Press, 1959.
157. YAGODA, H. *Radiation measurements with nuclear emulsions.* New York, Wiley; London, Chapman and Hall, 1949.

APPENDIX

PERIODICALS CONCERNED WITH ISOTOPES AND RADIATION AND THEIR APPLICATIONS IN INDUSTRY, SCIENCE AND MEDICINE

1. *Acta Isotopica.* A. Milani, via Japelli 5, Padua, Italy. 1961 and onwards. (Quarterly.)
   Original articles, surveys and technical notes on isotope applications in clinical and experimental medicine. Articles in Italian, English, French or German with abstracts in all four languages.

2. *Acta radiologica.* Box 2052, Stockholm, Sweden, 1921 and onwards. (Monthly.)
   Articles in English, French or German on medical radiology and nuclear medicine. Abstracts in all three languages.
3. AGRICULTURAL RESEARCH COUNCIL. RADIOBIOLOGICAL LABORATORY. *Annual report*, 1961/62. London, H.M.S.O. 1962 and onwards.
   Reports on the work of the Laboratory which is mainly concerned with the analysis of radioactivity in biological materials, including human food chains, and environmental contamination.
4. *American Journal of Roentgenology, Radium Therapy and Nuclear Medicine.* C. C. Thomas, 301—327 East Lawrence Avenue, Springfield, Illinois. 1913 and onwards. (Monthly.)
   Previously the *American Quarterly of Roentgenology*, 1906—13. Official organ of the American Roentgen Ray Society and the American Radium Society. Original articles on medical radiography and nuclear medicine. Each issue contains a section of abstracts of the radiological literature which are indexed, together with the articles, in the subject index issued for each volume of the journal.
5. *British Journal of Radiology.* British Institute of Radiology, 32 Welbeck St., London W.1. 1896 and onwards. (Monthly.)
   Includes articles on medical applications of nuclear science, particularly radiodiagnosis and radiotherapy.
6. *Bulletin d'Information sur les Applications Industrielles des Radioelements (Supplement to Bulletin d'Information de l'ATEN).* ATEN, 96 rue de Clichy, Paris 9e. 1963 and onwards. (Quarterly.)
   Short notes on developments in the applications of radioisotopes in industry with selected references to the current literature.
7. *Health Physics.* Pergamon Press, 122 West 55th Street, New York, 22, N.Y. 1958 and onwards. (Monthly.)
   Official organ of the Health Physics Society, Oak Ridge National Laboratory, Tennessee. Covers all aspects of protection against radiation, effects of radiations on man and the toxic effect of radiochemicals.
8. *International Journal of Applied Radiation and Isotopes.* Pergamon Press, Headington Hill Hall, Oxford. 1956 and onwards. (Monthly.)
   Articles and technical notes on isotope and radiation techniques and applications in industry, medicine and science with an international news section of general interest to isotope users.
9. *International Journal of Radiation Biology.* Taylor and Francis, Red Lion Court, Fleet Street, London, E.C.4. 1959 and onwards. (Monthly.)
   Original papers on the biological effects of radiations and related studies in physics, chemistry and medicine.
10. *Journal of Inorganic and Nuclear Chemistry.* Pergamon Press, Headington Hill Hall, Oxford. 1955 and onwards. (Monthly.)
    Articles and notes on inorganic chemistry, radiochemistry and radiation chemistry.
11. *Journal of Radiation Research.* Uchida's University Bookstore, 1—139 Totsukamachi, Shinjuku-ku, Tokyo. 1960 and onwards. (Irregular.)
    Organ of the Japan Radiation Research Society. Articles mainly on environmental contamination, including fall out, radiochemistry and radiobiology.
12. *Journal de Radiologie, d'Électrologie et de Médicine Nucléaire.* Masson et Cie, 120 boulevard St-Germain, Paris 6e. 1941 and onwards. (8 times a year.)
    Journal of the Société Française d'Électroradiologie Médicale et Filiales. Covers radioscopy, radiography, radiotherapy and electrology.
13. MEDICAL RESEARCH COUNCIL. *Report of the Medical Research Council.* London, H.M.S.O. (Annual.)

The report contains a general review of the Council's work during the year followed by a description of the research carried out by the various research units including the Radiobiological Research Unit, Harwell, Berks., and the Radiobiological Protection Service, Sutton, Surrey.

14. *Minerva Nucleare.* Edizioni Minerva Nucleare, corso Brimante 83/85, Turin, Italy. 1957 and onwards. (Bimonthly.)
Original articles, in Italian or English, on nuclear medicine, radiobiology, medical physics and health physics. A bibliography of current articles is included in each issue.

15. *Nuclear Medicine.* Excerpta Medica Foundation, 119—123 Herengracht, Amsterdam. January 1964 and onwards. (Monthly.)
Published in co-operation with Euratom, this is the main abstract journal in the field of nuclear medicine. Approximately 3000 abstracts p.a. on nuclear physics in biology and medicine, radiochemistry, health physics, radiobiology, isotope and radiation techniques in diagnosis and therapy, and treatment of radiation injury. There are author and subject indexes in each issue and these are cumulated annually.

16. *Physics in Medicine and Biology.* Taylor and Francis, Red Lion Court, Fleet Street, London, E.C.4. 1956 and onwards. (Quarterly.)
The official journal of the Hospital Physicists Association. Review articles and original papers on physical techniques used in biology and medicine including radiology, radiodiagnosis and radiotherapy. Each issue includes a section of abstracts on nuclear medicine, radiation dosimetry, radioisotopes and health physics but there is no annual index to these.

17. *Radiation Botany.* Pergamon Press, Headington Hill Hall, Oxford. 1961 and onwards. (Quarterly.)
Original articles, in English, French or German, on all aspects of plant radiobiology particularly the effects of radiation on plants.

18. *Radiation Research.* Academic Press, 111 Fifth Avenue, New York 3, N.Y. 1954 and onwards. (Monthly.)
Articles on the biological effects of all types of radiation including ultra-violet, infra-red and visible light. Covers a number of related subjects including dosimetry, isotope techniques, instrumentation and chemical agents contributing to the study of radiation effects.

19. *Radiochimica Acta.* Akademische Verlagsgesellschaft, Cronstetterstraße 6a, Frankfurt a. M., Germany *and* Academic Press, 111 Fifth Avenue, New York. 1962 and onwards. (Irregular.)
Original articles, in German, English or French, in the field of radiochemistry and radiochemical analysis. Abstracts in all three languages.

20. *Radioisotopes.* Japan Radioisotope Association, 31 Kamifujimaecho-Komagome, Bunkyo-ku, Tokyo. 1952 and onwards. (Bimonthly.)
Articles on biological and industrial application of radioisotopes, instrumental techniques, radiation protection and radiation injury. In Japanese with English abstracts.

21. *Radiological Health Data.* U.S. Government Printing Office, Washington 25, D.C. 1960 and onwards. (Monthly.)
Published by the Public Health Service of the U.S. Dept. of Health Education and Welfare. Includes data on environmental radiation levels (mainly in North America) and their interpretation.

22. *Radiology.* Radiological Society of North America, 20th and Northampton Streets, Easton, Pennsylvania. 1923 and onwards. Monthly.)
Original articles on clinical radiology and allied sciences.

23. *Strahlentherapie.* Urban & Schwarzenberg-Verlag für Medizin und Naturwissenschaften, Petternkofer Straße 18, Munich 15. 1912 and onwards. (Monthly.) Original papers, in German, on radiotherapy with abstracts in English and French.

CHAPTER 11

CONTROLLED NUCLEAR FUSION AND PLASMA PHYSICS

INTRODUCTION

The liberation of energy from the fusion of light elements was observed in experiments at the Cavendish Laboratory, Cambridge, as early as 1932 although the suggestion that such reactions might be used to generate useful power was described by Lord Rutherford, at the time, as "moonshine". For fusion to occur the nuclei must approach each other close enough for the short-range nuclear forces to interact, permitting the particles to coalesce. Unfortunately each of the reacting nuclei is positively charged, so that they strongly repel each other, and fusion can take place only if the nuclei are given sufficient kinetic energy to overcome this electrostatic repulsion. Clearly it is advantageous to use elements with the smallest possible nuclear charge such as hydrogen and its isotopes, deuterium and tritium. It is found that the most favourable reaction from the point of view of energy gain is that between the nuclei of deuterium and tritium; the gain is somewhat less for the reaction between two deuterium nuclei.

In order to endow deuterium nuclei with energies great enough to penetrate the electrostatic barrier, extremely high temperatures are required, of the order of tens of millions of degrees. Even so there is a greater probability that the nuclei will suffer elastic collisions than that they will fuse and they must therefore be confined in a region where they can be made to approach each other repeatedly until fusion takes place. At such high temperatures the atoms are stripped of their electrons and the resulting mixture of free electrons and positive ions in rapid random motion is known as a "plasma", a term introduced by Langmuir [24] in 1929, and now used to describe a highly ionized gas.

There are good reasons for believing that plasma is the normal state of matter in the universe. The sun and stars are sufficiently hot, and the interstellar matter is sufficiently diffuse, to be fully or partially ionized and only in planetary environments is plasma likely to be uncommon in comparison with the other three states of matter. Unfortunately physicists live and perform their experiments in a planetary environment in which it is by no means easy to produce a plasma or to preserve its existence long enough to study it adequately and the science of plasma physics is at a comparatively early stage of its development.

For a gas to be heated to fusion temperatures it must be confined and thermally insulated from its surroundings. Clearly, any material container is impracticable and gravitational forces, which confine the plasma in stellar bodies, are much too weak to be effective on a laboratory scale. Fortunately, since the plasma is electrically conducting, magnetic fields can be used to contain the plasma within a given region, provided that the field is strong enough to ensure that the inward magnetic pressure exerted by the field is greater than the outward pressure of the plasma. To keep the latter within reasonable bounds the density of the ionized gas must be low; about 1/10,000 of the density of the atmosphere is an average figure.

Even so magnetic confinement is far from perfect. A number of different kinds of instability develop and high energy particles penetrate the magnetic fields and escape. Energy is lost by radiation and by particles striking the sides of the containment vessels and the very high magnetic fields required can be sustained only for short time intervals. The problems are both theoretical and technological and are those of plasma physics and high voltage electrical engineering rather than nuclear physics. Thus although the end product, nuclear energy, is the same as in the case of nuclear fission, the research now being carried out in the fusion field has very little in common with atomic energy developments described in earlier chapters. It is, in fact, more closely related to studies in plasma propulsion, direct conversion and magnetohydrodynamic generation of power than to nuclear fission research.

ORGANIZATION OF FUSION RESEARCH

History

The possibility of using controlled nuclear fusion as a means of providing useful energy was discussed as early as 1946 in Britain and America, and possibly in other countries as well. In Britain experiments on the confinement of plasma by magnetic fields were started by Sir George Thomson in London in 1947 and by Dr. P. C. Thonemann at Oxford in 1948. In 1951 the work of the London group was transferred to the AEI Research Laboratories at Aldermaston and that of the Oxford group to Harwell. Later, in 1956, fusion experiments were begun at the Atomic Weapons Research Establishment at Aldermaston.

In 1959 the UKAEA began building a new laboratory at Culham, in Oxfordshire, to be devoted entirely to plasma physics and fusion studies, and the transfer of experimental work from Harwell and Aldermaston to Culham began in 1961 and is expected to be completed in 1965.

In the United States fusion studies were initiated in 1952 at Los Alamos Scientific Laboratory, at the University of California Lawrence Radiation Laboratory and at Princeton University. Later, in 1955, experiments were also begun at Oak Ridge National Laboratory and at the Naval Research Laboratory in Washington.

It is uncertain when fusion work started in the U.S.S.R. but in May 1956 Academician I. V. Kurchatov described in detail some of the Russian experiments, at a lecture given at Harwell. By the end of 1956 research was going on in a number of countries and in June 1957, at the Third International Conference on Ionization Phenomena in Gases, held in Venice, theoretical and experimental papers on fusion research were presented by Britain, the United States, the Soviet Union, France, Germany and Sweden, albeit in somewhat guarded terms.

In the spring of 1958 all fusion work was declassified by mutual consent and information began to flow freely between the various laboratories. It became apparent that research on parallel lines had been going on in a number of countries and at the second Geneva conference in September 1958 over 100 papers were presented on fusion studies.

United Kingdom

The centre for fusion research is the Culham Laboratory of the United Kingdom Atomic Energy Authority where a number of plasma containment systems are being studied. The work of the Laboratory has been described in a booklet, [1] which is revised periodically, and in the Annual Reports which are made available to other laboratories working on fusion research. Technical reports are issued and distributed in the same way as other UKAEA unclassified reports (see Chapter 2) and are listed in *N.S.A.* and the *UKAEA List of Publications Available to the Public*. In addition a number of internal reports, [2] including progress reports, are generated which are made available on an exchange basis to other fusion laboratories. Results of research are also published in papers submitted to scientific journals or presented at conferences and an annotated bibliography of the most important articles and conference papers has been issued. [3]

The work of the experimental team at the AEI Research Laboratory at Aldermaston came to an end in 1963 when that Laboratory was closed down and there is, at present, no British industrial organization carrying out research in the fusion field. Plasma physics studies are being conducted at Oxford University, Imperial College, the University College of Wales at Aberystwyth and at the New University of Sussex at Brighton. Research into natural plasmas is part of the programme of the Radio Research Station at Slough. [4]

United States

The USAEC programme of fusion research, known as Project Sherwood, is an integrated programme in which five laboratories are taking part, namely, the Los Alamos Scientific Laboratory, the University of California Lawrence Radiation Laboratory at Berkeley and Livermore, Princeton University and Oak Ridge National Laboratory. All these publish technical reports which are distributed in the same manner as other AEC documents (see Chapter 3), and the Divisions concerned with thermonuclear work issue progress reports [5, 6, 7, 8] which are publicly available. The Lawrence Radiation Laboratory also publishes an annual status report [9] from which a broad picture of fusion studies in the Laboratory can be obtained.

The Plasma Physics Laboratory at Princeton University, where fusion work is conducted under the code name "Project Matterhorn", has issued a cumulated list of its reports and papers published up to 1961, [10] which is supplemented by the list of publications in each copy of the progress report. Reports issued by the Lawrence Radiation Laboratory up to 1958 are listed in a bibliography [11] issued in 1959 and this too is kept up to date by lists in the progress reports of the Thermonuclear Division.

Fusion research is undertaken by a number of laboratories outside the AEC. For example, fusion studies have been going on since 1955 in the Radiation Division of the U.S. Naval Research Laboratory, Washington; the reports and papers issued are listed in the monthly *Report of NRL Progress.* [12] Of the industrial organizations working in this field, Boeing Scientific Research Laboratories, Seattle, publish a half-yearly progress review, [13] and technical reports are issued by General Atomic Division of General Dynamics Corporation, San Diego, the Avco-Everett Research Laboratory at Everett, Massachusetts, and the General Electric Research Laboratory at Schenectady, New York. The latter issues a quarterly *Bulletin* [14] which contains occasional notes on the fusion work of the Laboratory.

Soviet Union

Fusion studies are carried on in a number of laboratories including the I. V. Kurchatov Institute of Atomic Energy, Moscow, the Physico-Technical Institute, Kharkov and the Lebedev Institute for Physics in Moscow, all of which issue reports and preprints. Experiments are also in progress at the Electro-Physical Institute and the Efremov Scientific Research Institute for Electrical and Physical Equipment, both of which are in Leningrad, the Institute for Physics and Engineering at Sukhumi and the Institute of Nuclear Physics at Novosebirsk.

In general the work of these laboratories is reported in *Zhurnal Eksperimental'noi i Teoreticheskoi Fiziki, Zhurnal Technicheskoi Fiziki,* and other Russian physics journals (see Chapter 7 for translated versions), and few reports are issued.

Euratom

In Europe there are five major laboratories engaged in fusion research, all of them operating under contracts of association with Euratom. A brief review of the fusion programme of these laboratories appeared in *Atomwirtschaft* [15] in 1963.

The Euratom—CEA programme which began in 1959 is based on the French nuclear centre at Fontenay-aux-Roses and details of the experimental work are published in the annual progress report, [16] each issue of which contains a list of reports and papers published during the year. A cumulated list [17] of reports and papers issued by the Fusion Group at Fontenay during the period 1958 to June 1964 has also been issued.

In 1961 Euratom entered into agreements with the Italian CNEN, to embrace the fusion studies at the Laboratorio Gaz Ionizzati at Frascati, and with the Institut für Plasmaphysik, Garching bei Munchen, which is probably the largest fusion laboratory in Europe. The Institut works in co-operation with the Max-Planck-Institut für Physik und Astrophysik in Munich and the two organizations issue a combined progress report [18] which is circulated to other plasma physics laboratories and contains a list of published reports and papers.

In 1962 an agreement was made with KFA (Kernforschungsanlage Jülich des Landes Nordrhein-Westfalen) to include the fusion programme of the Institut für Plasmaphysik at Jülich. This Institute issues an annual progress report [19] which is given a limited circulation. In the same year a similar contract was made with the Jutphaas Laboratory of the Dutch Foundation for Fundamental Research on Matter (FOM).

Plasma physics studies, not necessarily connected with fusion research, are being undertaken in a number of other laboratories in Euratom countries and particularly in Western Germany. A brief description of some of the German laboratories has appeared in *Atomwirtschaft*. [20]

Other Countries

Apart from the ones already mentioned there are a number of smaller fusion projects in Europe, Australia and Japan. In Europe work is in progress at the Institute of Plasma Physics of the Czechoslovak Academy of Sciences in Prague, the Risø Research Centre in Denmark, the Institute of Physics of the University of Bergen in Norway, the University of Uppsala and the Royal Institute of Technology, Stockholm, in Sweden and the Laboratoire de Recherches sur la Physique des Plasmas in Lausanne, Switzerland. Most of these issue technical reports which are made available to other organizations in the field.

In Australia the main centre is the Wills Plasma Physics Laboratory of the University of Sydney, which issues a half-yearly progress report, [21] and plasma physics research is also done at Adelaide University and the National University of Australia at Canberra.

Fusion research in Japan is undertaken by the Institute of Plasma Physics of Nagoya University, which issues reports and an *Annual Review* [22] of the work of the Nuclear Fusion Group. The latter publishes *Kakuyugo Kenkyu*, [23] a monthly journal in Japanese which contains original papers and lists recently published reports and papers originating in Japan and elsewhere. The Institute houses the Research Information Centre which provides a national information service in this field and is the main depository for plasma physics literature in Japan.

PUBLISHED INFORMATION
Books
Gas Discharges

Plasma physics research, in common with research in atomic and nuclear physics, originated largely in the early studies of the conduction of electricity through gases, although for some twenty years after Langmuir's [24] pioneer work in the nineteen-twenties, little interest was displayed in plasma phenomena. However a study of gas discharge processes is a necessary foundation for an understanding of plasma physics and one of the earliest modern texts is Cobine's *Gaseous Conductors* [25] which describes the basic phenomena in electrical discharges in gases and refers extensively to earlier work on the subject. Ionization processes in gases are discussed in detail in a standard work by Von Engel, [26] which emphasizes the physical understanding of the mechanisms involved, and in two advanced texts by Llewellyn-Jones [27] and Loeb, [28] which pay particular attention to the phenomena occurring during electrical breakdown. Of more immediate interest to the physicist working on nuclear fusion is *Ionization Phenomena in Gases* [29] which avoids discussion of the more familiar arc, spark and glow discharges and treats those aspects which are linked with other branches of physics and particularly high current discharges and thermonuclear research.

Plasma Physics

If there is anything approaching a "bible" in plasma physics it is undoubtedly Spitzer's *Physics of Fully Ionized Gases,* [30] the first edition of which was, for many years, the only modern book on plasmas. This theoretical introduction emphasizes collective as opposed to single-particle behaviour in the plasma and was written at a time when the author was already engaged in fusion studies. In contrast, Delcroix's *Introduction to the Theory of Ionized Gases* [31] adopts the single-particle approach, describing the non-collective behaviour of ions and electrons in magnetic fields.

Probably the best introduction to the subject for those entering the fusion field is *Plasmas and Controlled Fusion,* [32] by Rose and Clark, which starts at a fairly elementary level but rapidly increases in complexity. This is both a text book for the student and a reference book for the specialist, although the bibliography is not extensive. A more sophisticated theoretical treatment is given by Thompson [33] who is concerned chiefly with an idealized plasma in which collision processes are relatively unimportant. The argument is developed in the context of fusion research although experiments are described only in so far as they have some bearing on theoretical prediction. A similar introduction by Longmire [34] adopts a mainly mathematical approach and is again written primarily with fusion studies in mind, the examples being drawn from high-temperature, low-density plasmas. This book contains a long bibliography of original papers. Uman's *Introduction to Plasma Physics* [35] is a short course in the fundamentals of the subject for engineers and physicists in industry. A less mathematical introduction is Gartenhaus's *Elements of Plasma Physics,* [36] a theoretical account for graduate students. In a somewhat different class is Linhart's *Plasma Physics,* [37] which is a review of fundamental principles, with chapters on the main applications including thermonuclear research, direct conversion, rocket propulsion and energy storage. A long bibliography to important papers is included.

The theoretical study of the properties of highly ionized gases may be approached either through the microscopic description of the interactions between individual particles or through a macroscopic description in which the plasma is considered as a conducting fluid which obeys the laws of magnetohydrodynamics. In the first case the starting point is an analysis of the collision processes which govern the distribution of the various particles in the gas and a comprehensive account of these is given by Massey and Burhop [38] in what has become a standard work on impact phenomena. Brown's *Basic Data of Plasma Physics* [39] is also largely concerned with the motion of charged particles and their interactions. A more recent work is Hasted's *Physics of Atomic Collisions* [40] which gives a comprehensive account of all collision phenomena, including those in ionized gases, and provides a complete groundwork for the experimental physicist entering this field.

For the macroscopic approach a starting point is Cowling's *Magnetohydrodynamics* [41] which studies the basic physics of an electrically conducting fluid in a magnetic field and is mainly concerned with geophysical and astrophysical aspects of the subject. *An Introduction to Magnetofluidmechanics,* [42] by Ferraro and Plumpton, is also based on astrophysical studies and covers a wide range of topics in plasma physics. The material is derived largely from a lecture series given by one of the authors in 1958, so that references to fusion experiments are almost entirely lacking, but the book is notable for an exceptionally clear exposition of magnetohydrodynamics. A review of magnetohydrodynamic theory forms part of Drummond's *Plasma Physics,* [43] a collection of essays, by several contributors, on various aspects of plasma research, including plasma dynamics and plasma oscillations. A similar volume is Clauser's *Symposium on Plasma Dynamics* [44] which presents a broad picture of the field and has a long bibliography of over 70 pages.

Plasma Dynamics

Of particular importance in controlled fusion research are the various forms of instabilities which occur in a plasma. Macroscopic instabilities are discussed by Chandrasekhar [45] in a particularly lucid account of a subject which bristles with difficulties and in a collection of Russian papers, [46] most of them by scientists working on controlled fusion, which is in course of translation by Consultants Bureau. Micro-instabilities and the various kinds of small-amplitude oscillations which have been detected in plasmas are discussed at some length in the literature. A classification of the known types of plasma waves has been attempted by Stix [47] in an elegant treatment which is concerned mainly with the low-density, high temperature plasmas of interest in fusion research. A more detailed analysis is given by Allis, Buchsbaum and Bers [48] in a mathematical account of free and guided wave propagation in plasmas. This book has a long bibliography which includes many references to Russian work in this field. Other mathematical treatments are those of Denisse and Delcroix [49] and Brandstatter, [50] the latter introducing material on the interaction between electromagnetic radiation and plasmas.

The propagation of electromagnetic waves in plasmas was first studied in connection with radio research and particularly the effect of the ionosphere on radio waves. A major work on this subject is Ratcliffe's *Magneto-ionic Theory* [51] which is a comprehensive account of the interaction between electromagnetic radiation and a partially-ionized gas. There has been some excellent Russian work on this subject, much of which is summarized in Ginzburg's *Propagation of Electromagnetic Waves in Plasma,* [52] which also emphasizes the geophysical and cosmic aspects of the subject. In fact, plasma physics owes a great deal to the astrophysicists' earlier studies of natural plasmas in the stars and in interplanetary and interstellar space, studies which have been described in the first edition of Alfen's *Cosmical Electrodynamics.* The second edition [53] of this book is a very much expanded version of the first four chapters of the earlier edition, dealing only with the fundamental principles of plasma physics and omitting any more than passing reference to the astrophysical aspects of the subject. Dungey's book [54] with a similar title is a straightforward introduction to plasma physics and magnetohydrodynamics with the emphasis on natural rather than laboratory plasmas. Another astrophysical reference book which is of great value in plasma physics studies is Allen's *Astrophysical Quantities,* [55] a fully documented tabulation of the essential quantitative data of astrophysics.

Controlled Thermonuclear Research

The books referred to so far are essentially theoretical treatments of the fundamental physics upon which experiments in controlled fusion are based. Although the literature of fusion research is extensive, a good deal of it is in the form of reports, journal articles, and conference papers. For example, one of the most readable summaries [56] of the problems to be overcome in developing a fusion reactor appeared in *Reviews of Modern Physics* in 1956, although no description of the actual experimental work was possible at that time for security reasons. A similar review of possible approaches to controlled fusion is the subject of Simon's book [57] which is based on a series of lectures given in 1955 and is now somewhat out of date, although the chapter on plasma stability is still valuable.

The first comprehensive account of experimental work in this field is Bishop's *Project Sherwood*, [58] written specially for the *Atoms for Peace* series of volumes which was presented by the USAEC to other official atomic energy projects at the time of the second Geneva conference in 1958. This is a fairly elementary description of fusion research in the United States between 1951 and 1958, well illustrated and complete with a bibliography of selected references. A number of original papers on fusion and plasma physics were presented at Geneva and the most useful of these are published in *Nuclear Fusion* [59] in an edited form. Russian experiments during 1951–1958 were also described at Geneva in a series of papers which appeared in the proceedings of the conference. Other Russian papers describing the theoretical and experimental work carried out during this period, mainly at the Atomic Energy Institute in Moscow, have been collected and translated in a four volume work [60] edited by Leontovitch. Only papers which had not previously appeared in print are included. A concise review of fusion research, which pays particular attention to developments since the Geneva conference, is contained in an IAEA booklet, [61] issued in the Review Series in 1961. Over 200 references to original papers are given together with a list of the most useful bibliographies in this field.

The most complete introduction to both theoretical and experimental aspects of fusion research, at intermediate level, is Glasstone and Lovberg's *Controlled Thermonuclear Reactions*, [62] which is mainly an account of American experience, with few references to experiments in other countries. Nevertheless for a library that wished to acquire one representative book on fusion research, this would be the best choice. British experiments, together with those in other countries, are described in Saxe's *Approaches to Thermonuclear Power*, [63] a concise summary of basic principles and the main lines of research for nuclear engineering students and others who require a broad understanding of the subject. The most up-to-date introduction to experimental work on fusion is Green's *Thermonuclear Power* [64] which describes the main experimental approaches and discusses the theoretical problems and the principal methods of making measurements of plasma parameters. This is a practical textbook for scientists and engineers entering the field and has a selected list of references for further reading.

Many of the books for non-technical readers mentioned in Chapters 7 and 9 contain chapters on thermonuclear research, although in most cases the subject is dealt with in rather general terms. A more detailed introduction, for the layman, to the basic ideas of fusion research is Jukes's *Man-made Sun*, [65] in which the principal problems are illustrated by reference to the ZETA experiment. A more broadly-based account of fusion experiments is contained in a booklet [66] issued by the University of California, Lawrence Radiation Laboratory, and written by one of America's foremost scientists in the fusion field. Slightly more advanced but still within the understanding of the educated reader is Étievant's *L'Énergie Thermonucléaire* [67] which discussed basic principles and the main experimental developments in the major countries. Lastly, a book which introduces the sociological aspects of thermonuclear power, is *The Challenge of Fusion* [68] by Curry and Newman, in which the first part provides a simple introduction to the scientific problem of controlled fusion and the second part considers the implications for society of a readily available source of unlimited energy.

Conference Proceedings

Reference has already been made to the second International Conference on the Peaceful Uses of Atomic Energy, held in Geneva in 1958, at which a great deal of information on fusion research was made available for the first time. Three other series of conferences are important in this field. The International Conference on Ionization Phenomena in Gases is held approximately every two years and although one or two rather theoretical papers on fusion were included in the proceedings of the third conference, held in Venice in 1957, the first meeting of importance from the point of view of fusion was the fourth conference held in Uppsala in 1959. [69] The fifth [70] and sixth [71] conferences included papers on gas discharges and plasma physics, but little on experimental work on controlled fusion because, in 1961, the International Atomic Energy Agency organized the first large-scale international conference devoted exclusively to fusion research. The proceedings, [72] which were published as a special supplement to the journal *Nuclear Fusion*, are a rich source of information on the latest developments in experimental and theoretical work in this field.

An important series of conferences is the Lockheed Symposia on Magnetohydrodynamics which are held annually, although not all of these meetings are of interest to fusion workers. The papers are mainly theoretical studies on plasma physics and those having most bearing on controlled thermonuclear research are contained in the proceedings of the second, [73] fifth, [74] sixth, [75] and seventh [76] symposia.

Bibliographies

Apart from those already mentioned in the publications described above, there are a number of separate bibliographies available. The most comprehensive is that published by the IAEA [77] in its bibliography series. This contains over 5000 references to reports and articles in English, Russian, French, German, Italian, Japanese and other languages, published between 1955 and 1961. The references are classified under twenty broad headings covering plasma physics, stability studies, magnetohydrodynamics, shock waves, fusion experiments and measurement techniques, and, for Western readers, the only thing which detracts from its usefulness is that many of the references and abstracts are in Russian.

A series of bibliographies issued by the USAEC [78, 79] lists reports and articles which were abstracted in Nuclear Science Abstracts from 1952 onwards. Fusion studies in the United States and, to a lesser extent, in other countries are covered selectively and each issue contains an availability index to the reports listed. British reports and papers, published up to September 1960, are listed, without abstracts, in a UKAEA bibliography [80] issued in 1960. Another UKAEA bibliography [81] contains references to all important articles and reports dealing with the interaction between electromagnetic radiation and plasmas, and particularly radio-frequency containment of plasmas. A more recent bibliography [82] lists publications on fusion devices and experiments which employ a "magnetic mirror" geometry for containment of the plasma.

It is worth mentioning two bibliographies on plasma physics and magnetohydrodynamics which, although not immediately concerned with fusion research, contain a good deal of fringe material. These are the long bibliography compiled by ASTIA [83] (now the Defence Documentation Center), which includes references on pinch studies and plasma instabilities, and a comprehensive bibliography on magnetohydrodynamics, [84] covering the period from 1920 to 1960, which has been issued in the AGARD Bibliography series.

REFERENCES

Organizations

1. UNITED KINGDOM ATOMIC ENERGY AUTHORITY. *Culham Laboratory for plasma physics and fusion research.* UKAEA Culham Laboratory, Abingdon, Berks., 1965.
2. UKAEA CULHAM LABORATORY. *Reports, etc. issued up to June, 1964.* Library, Culham Laboratory, Abingdon, Berks., 1964.
3. UKAEA CULHAM LABORATORY. *An annotated bibliography of articles on plasma physics and controlled thermonuclear research by UKAEA staff, 1958 to 1962.* London, H.M.S.O., 1963. (CLM—R 24.)
4. DEPARTMENT OF SCIENTIFIC AND INDUSTRIAL RESEARCH. *Radio Research: report of the Radio Research Board and the report of the Director of Radio Research.* London, H.M.S.O., 1928/29 and onwards.
   Includes lists of papers and reports issued.
5. LOS ALAMOS SCIENTIFIC LABORATORY. *Semiannual status report of the LASL controlled thermonuclear research program.* Washington, Office of Technical Services, June 1963 and onwards.
   Includes lists of reports and papers issued during each six-monthly period. From March 1960 to May 1963 quarterly progress reports were issued.
6. PRINCETON UNIVERSITY. PLASMA PHYSICS LABORATORY. *Annual report.* Washington, Office of Technical Services, January 1963 and onwards.
   From October 1957 to June 1962 a quarterly progress report was issued followed by one semi-annual report covering July—December 1962.
7. UNIVERSITY OF CALIFORNIA. LAWRENCE RADIATION LABORATORY. *Controlled thermonuclear research semiannual report.* Washington, Office of Technical Services, July 1961 and onwards.
   From December 1958 to June 1961 a quarterly progress report was issued.
8. OAK RIDGE NATIONAL LABORATORY. *Thermonuclear Division semiannual progress report.* Washington, Office of Technical Services, 1957 and onwards.

9. UNIVERSITY OF CALIFORNIA. LAWRENCE RADIATION LABORATORY AT BERKELEY AND LIVERMORE. *Status report*, 1960/61 and onwards. Berkeley, California, Lawrence Radiation Laboratory.
10. PRINCETON UNIVERSITY. PLASMA PHYSICS LABORATORY. *Cumulative listing of Plasma Physics Laboratory reports and publications. (July 1951—November 1961).* Princeton, N.J., James Forrestal Research Centre, November 1961. (MATT—1, 3rd ed.)
11. UNIVERSITY OF CALIFORNIA. LAWRENCE RADIATION LABORATORY. *Bibliography—controlled thermonuclear processes. LRL Berkeley, 1952—1958; LRL Livermore, 1953—1958.* Washington, Office of Technical Services, 1959. (UCRL 9019.)
12. U.S. NAVAL RESEARCH LABORATORY. *Report of NRL progress.* Washington, Office of Technical Services. (Monthly.)
13. BOEING SCIENTIFIC RESEARCH LABORATORIES. PLASMA PHYSICS LABORATORY. *Progress review.* (Half-yearly.) Seattle, Boeing Scientific Laboratories, P.O. Box 3981. 1958 and onwards.
14. GENERAL ELECTRIC COMPANY. *Research Laboratory Bulletin.* General Electric Co., Schenectady, New York. (Quarterly.) See particularly the issue for Summer 1964.
15. PALUMBO, D. *Das Euratom Program zur Fusionsforschung.* Atomwirtschaft, vol. 8, no. 5, pp. 297—300, May 1963. (Also issued as Euratom Reprint 442 d.)
16. EURATOM-CEA. GROUPE DE RECHERCHES SUR LA FUSION. *Rapport d'Activité du Group de Recherches.* Centre d'Études Nucléaires du Fontenay-aux-Roses, Boite Postale No. 6, Fontenay-aux-Roses (Seine), 1963 and onwards.
17. EURATOM—CEA. GROUP DE RECHERCHES SUR LA FUSION. *Liste des publications.* Centre d'Études Nucléaires de Fontenay-aux-Roses, Boite Postale No. 6, Fontenay-aux-Roses (Seine), 1964.
18. INSTITUT FÜR PLASMAPHYSIK, GMBH. *Jahresbericht des Instituts für Plasmaphysik GmbH, München-Garching und der Experimentellen Abteilung (Plasmaphysik) des Max-Planck-Instituts für Physik und Astrophysik, München.* Institut für Plasmaphysik GmbH, Garching bei München, 1962 and onwards.
19. EURATOM—KFA. INSTITUT FÜR PLASMAPHYSIK DER KERNFORSCHUNGSANLAGE JULICH. *Arbeitsberichte der theoretischen und experimentellen Gruppen.* Institut für Plasmaphysik, Julich, 1963 and onwards.
20. BARTELS, H. Die Forschungen auf dem Gebiet der Plasmaphysik in der Bundesrepublik Deutschland. *Atomwirtschaft*, vol. 8, no. 5, pp. 294—297, May 1963.
21. UNIVERSITY OF SYDNEY. SCHOOL OF PHYSICS. WILLS PLASMA PHYSICS DEPT. *Six-monthly progress report.* The Department, January 1961 and onwards.
22. NAGOYA UNIVERSITY. INSTITUTE OF PLASMA PHYSICS. *Annual review.* (In English.) Institute of Plasma Physics, Chikusa-Ku, Nagoya, 1961 and onwards. (First issue covered April 1961 to March 1963).
23. *Kakuyugo Kenkyo.* (Monthly.) Nuclear Fusion Group, Institute of Plasma Physics, Nagoya University, Nagoya 1958 and onwards. (In Japanese with English abstracts.)

Books

24. LANGMUIR, I. *The collected works of Irving Langmuir; with contributions in memoriam including a complete bibliography of his works*, edited by C. G. Suits and H. E. Way, Oxford, Pergamon Press, 1961.
    The relevant volumes are vol. 4—Electrical discharges, vol. 5—Plasma oscillations.
25. COBINE, J. D. *Gaseous conductors: theory and engineering applications.* New York, Dover; London, Constable, 1958. (Reprint of 1941 edition.)
26. VON ENGEL, A. *Ionized gases.* London, O.U.P., 1955.

27. LLEWELLYN-JONES, F. *Ionization and breakdown in gases.* London, Methuen; New York, Wiley, 1957.
28. LOEB, L. B. *Basic processes of gaseous electronics.* 2nd ed. rev. Berkeley and Los Angeles, University of California Press, 1961.
29. FRANCIS, G. *Ionization phenomena in gases.* London, Butterworth, 1960.
30. SPITZER, L. *Physics of fully ionized gases.* 2nd rev. ed. New York, London, Interscience, 1962.
31. DELCROIX, J. L. *Introduction to the theory of ionized gases,* translated by M. Clark and others. New York, London, Interscience, 1960. (First published in Paris, 1959.)
32. ROSE, D. J. AND CLARK, M. *Plasmas and controlled fusion.* Cambridge (Mass.), M.I.T. Press and New York, Wiley, 1961.
33. THOMPSON, W. B. *An introduction to plasma physics.* Oxford, Pergamon Press, 1962.
34. LONGMIRE, C. L. *Elementary plasma physics.* New York, London, Interscience, 1963.
35. UMAN, M. A. *Introduction to plasma physics.* New York, McGraw-Hill, 1964.
36. GARTENHAUS, S. *Elements of plasma physics.* New York, Holt, Rinehart and Winston, 1964.
37. LINHART, J. G. *Plasma Physics.* 2nd ed. Amsterdam, North Holland Pub. Co., 1961.
38. MASSEY, H. S. W. AND BURHOP, E. H. S. *Electronic and ionic impact phenomena.* Oxford, Clarendon Press, 1952.
39. BROWN, S. C. *Basic data of plasma physics.* New York, Wiley; London, Chapman and Hall, 1959.
40. HASTED, J. B. *Physics of atomic collisions.* London, Butterworth, 1964.
41. COWLING, T. G. *Magnetohydrodynamics.* New York, London, Interscience, 1957.
42. FERRARO, V. C. A. AND PLUMPTON, C. *An introduction to magnetofluidmechanics.* London, O.U.P., 1961.
43. DRUMMOND, J. E. *Plasma physics.* New York, McGraw-Hill, 1961.
44. CLAUSER, F. H. *Symposium on plasma dynamics.* Reading (Mass.), Addison Wesley; London, Pergamon Press, 1960.
45. CHANDRASEKHAR, S. *Hydrodynamic and hydromagnetic stability.* London, O.U.P., 1961.
46. LEONTOVICH, M. A., ED. *Problems in plasma theory.* 2nd ed. Moscow, Gosatomizdat, 1963. 3 vols.
    In course of translation by Consultants Bureau.
47. STIX, T. H. *Theory of plasma waves.* New York, McGraw-Hill, 1962.
48. ALLIS, W. P. AND OTHERS. *Waves in anisotropic plasmas.* Cambridge (Mass.), M.I.T. Press, 1963.
49. DENISSE, J. F. AND DELCROIX, J. L. *Plasma waves,* translated by M. Weinrach and D. J. BenDaniel. New York, London, Interscience, 1963. (First published in Paris in 1961.)
50. BRANDSTATTER, J. J. *An introduction to waves, rays and radiation in plasma media.* New York, McGraw-Hill, 1963.
51. RATCLIFFE, J. A. *The magneto-ionic theory and its application to the ionosphere.* London, C. U. P., 1959.
52. GINZBURG, V. L. *Propagation of electromagnetic waves in plasma,* translated from the Russian by J. B. Sykes and R. J. Tayler. Oxford, Pergamon Press, 1964. (First published in Moscow, 1960.)
53. ALFVEN, H. AND FÖLTHAMMAR, C. *Cosmical electrodynamics: fundamental principles.* 2nd ed. Oxford, Clarendon Press, 1963. (First edition, 1960.)
54. DUNGEY, J. W. *Cosmic electrodynamics.* London, C.U.P., 1958.

55. ALLEN, C. W. *Astrophysical quantities.* 2nd ed. University of London, Athlone Press, 1963.
56. POST, R. F. Controlled fusion research: an application of the physics of high temperature plasmas. *Rev. Mod. Phys.,* vol. 28, no. 3, pp. 338—362, July 1956.
57. SIMON, A. *An introduction to thermonuclear research.* London, Pergamon Press, 1959.
58. BISHOP, A. S. *Project Sherwood: an account of the program in controlled thermonuclear reactions carried out by the USAEC during the period 1951—1958.* Reading (Mass.), Addison-Wesley, 1958.
59. ALLIS, W. P., ED. *Nuclear fusion.* Princeton, Van Nostrand, 1960.
60. LEONTOVICH, M. A., ED. *Plasma physics and the problem of controlled thermonuclear reactions.* (In English.) Oxford, Pergamon Press, 1959—1961. 4 vols.
61. BOCKASTON, K. AND OTHERS. *Controlled thermonuclear fusion research.* IAEA, Vienna 1, 1961. (Review series no. 17.)
62. GLASSTONE, S. AND LOVBERG, R. H. *Controlled thermonuclear reactions: an introduction to theory and experiment.* Princeton, Van Nostrand, 1960.
63. SAXE, R. F. *Approaches to thermonuclear power.* London, Temple Press, 1960.
64. GREEN, T. S. *Thermonuclear power.* London, Newnes, 1963.

Books for the General Reader

65. JUKES, J. D. *Man-made sun: the story of ZETA.* London, New York, Abelard-Schuman, 1959.
66. POST, R. F. *The fourth state of matter—plasma.* University of California, Lawrence Radiation Laboratory, Livermore, California, 1962.
67. ÉTIEVANT, C. *L'Énergie thermonucléaire.* Paris, Presses Universitaires de France, 1962.
68. CURRY, D. AND NEWMAN, B. R. *The challenge of fusion.* Princeton, Van Nostrand, 1960.

Conferences

69. FOURTH INTERNATIONAL CONFERENCE ON IONIZATION PHENOMENA IN GASES, UPPSALA, AUGUST 1959. *Proceedings,* ed. by N. R. Nilsson. Amsterdam, North Holland, 1960. 2 vols.
70. FIFTH INTERNATIONAL CONFERENCE ON IONIZATION PHENOMENA IN GASES, MUNICH, AUGUST—SEPTEMBER, 1961. *Proceedings,* ed. by H. Maecker. Amsterdam, North Holland, 1962. 2 vols.
71. SIXTH INTERNATIONAL CONFERENCE ON IONIZATION PHENOMENA IN GASES, PARIS, JULY, 1963. *Proceedings,* ed. by P. Hubert and E. Cremieu-Alcan. Orsay (S. & O.) France, Conférence Internationale sur les Phénomenes d'Ionization dans les Gaz, Faculté des Sciences, Boite Postale No. 2, 1964. 4 vols.
72. INTERNATIONAL ATOMIC ENERGY AGENCY. *Proceedings of the Conference on Plasma Physics and Controlled Nuclear Fusion Research, Salzburg, September, 1961.* Nuclear Fusion 1962 Supplement, Parts 1—3, 1962—1963.
73. LANDSHOFF, R. K. M., ED. *The plasma in a magnetic field: proceedings of the 2nd Lockheed Symposium on Magnetohydrodynamics.* Stanford University Press, 1958.
74. MITCHNER, M., ED. *Radiation and waves in plasmas: proceedings of the 5th Lockheed Symposium on Magnetohydrodynamics.* Stanford University Press, 1961.
75. BERSHADER, D., ED. *Plasma hydromagnetics: proceedings of the 6th Lockheed Symposium on Magnetohydrodynamics.* Stanford University Press, 1962.

76. FUTTERMAN, W. I., ED. *Propagation and instabilities in plasmas: proceedings of the 7th Lockheed Symposium on Magnetohydrodynamics.* Stanford University Press, 1963.

Bibliographies

77. INTERNATIONAL ATOMIC ENERGY AGENCY. *Research on controlled thermonuclear fusion.* Vienna, I.A.E.A., 1962. (Bibliographical series no. 7.)
78. (a) UNITED STATES ATOMIC ENERGY COMMISSION. OFFICE OF TECHNICAL INFORMATION. *Controlled thermonuclear processes: a selective bibliography.* Washington, Office of Technical Services, May 1958. (TID—3072.)
Abstracts of unclassified reports and journal articles up to May 1958.
(b) UNITED STATES ATOMIC ENERGY COMMISSION. OFFICE OF TECHNICAL INFORMATION, *Controlled thermonuclear reactions: a selected bibliography*, compiled by S. F. Lanier, R. L. Scott and T. W. Scott. Washington, Office of Technical Services, January 1961. (TID—3072, rev. 1.)
Abstracts of reports and articles issued May 1958 to December 1959 including those Project Sherwood reports declassified in Autumn 1958.
79. (a) UNITED STATES ATOMIC ENERGY COMMISSION. OFFICE OF TECHNICAL INFORMATION. *Controlled thermonuclear processes: a literature search*, compiled by R. L. Scott and S. F. Lanier. Washington, Office of Technical Services, November 1960. (TID—3557.)
No abstracts. Covers reports and papers issued December 1959 to December 1960. All entries are taken from *N.S.A.* and each refers to the *N.S.A.* abstract number.
(b) UNITED STATES ATOMIC ENERGY COMMISSION. DIVISION OF TECHNICAL INFORMATION. *Controlled fusion and plasma research: a literature search*, compiled by S. F. Lanier. Washington, Office of Technical Services, August 1962. (TID—3557, suppl. 1.)
No abstracts. Covers January 1961 to May 1962. All entries are taken from *N.S.A.* and each refers to the *N.S.A.* abstract number.
80. SABEL, C. S., COMP. *UKAEA and associated British work on controlled thermonuclear reactions: a list of unclassified documents and published articles.* London, H.M.S.O., 1960. (AERE-Bib. 124, rev. 1.)
81. KNOX, F. B. AND SABEL, C. S. *Oscillating electromagnetic field interaction with, and containment of, plasmas.* London, H.M.S.O., 1959. (AERE-Bib. 121.)
First supplement published as CLM-Bib. 1, H.M.S.O., 1961; second supplement published as CLM-Bib. 4, H.M.S.O., 1964.
82. FRANCIS, L. S., COMP. *Magnetic mirror machines: a bibliography.* London, H.M.S.O., 1964. (CLM-Bib. 5.)
Covers period from 1954 to early 1963.
83. (a) ARMED SERVICES TECHNICAL INFORMATION AGENCY. *Plasma physics and magnetohydrodynamics: an ASTIA report bibliography,* compiled by M. F. Aukland. Washington, Office of Technical Services, March 1962. (AD 271170.)
Report issued from 1953 to early 1962.
(b) DEFENSE DOCUMENTATION CENTER. *Plasma physics and magnetohydrodynamics: a report bibliography,* compiled by M. F. Aukland. Defense Documentation Center, Cameron Station, Alexandria, Virginia, June 1963. (AD 405732.)
Supplement to AD 271170.
84. NAPOLITANO, L. G. AND CONTURSI, G. *Magnetofluiddynamics: current papers and abstracts.* Oxford, Pergamon Press, 1962. (AGARD Bibliography No. 1, 2nd ed.)
Over 2000 entries, of which 700 carry abstracts, covering 1940 to 1960.

Appendix

Periodicals Containing Articles on Plasma Physics and Controlled Fusion Research

1. *Beiträge aus der Plasmaphysik.* Akademie Verlag G.m.b.H., Leipziger Straße 3, 108 Berlin. 1960 and onwards. (Irregular.)
   Original articles on plasma physics and magnetohydrodynamics reporting research in East Germany, Czechoslovakia, etc.
2. *Journal of Fluid Mechanics.* Cambridge University Press, Bentley House, 200 Euston Road, London, N.W.1. 1956 and onwards. (Monthly.)
   Occasional articles on plasma dynamics and magnetohydrodynamics.
3. *Journal of the Physical Society of Japan.* Physical Society of Japan, Physics Building, Faculty of Science, University of Tokyo, Bunkyo-ku, Tokyo. 1946 and onwards. (Monthly.)
   Original articles, in English, on all aspects of physics but with a high proportion of papers reporting plasma physics research in Japan.
4. *Journal of Quantitative Spectroscopy and Radiative Transfer.* Pergamon Press, Headington Hill Hall, Oxford. 1961 and onwards. (Bimonthly.)
   Occasional articles on spectroscopic measurements of plasma parameters.
5. *Nuclear Fusion.* International Atomic Energy Agency, Kärntner Ring 11, Vienna 1. 1960 and onwards. (Quarterly.)
   Original and review articles on plasma physics and controlled fusion research. Articles in English, Russian, French or Spanish with abstracts in all four languages. Each issue has a section devoted to notices of forthcoming meetings and conferences on plasma physics and associated subjects.
6. *Physics of Fluids.* American Institute of Physics, 335 East 45 St., New York, N.Y. 1958 and onwards. (Monthly.)
   Although ostensibly devoted to all aspects of the physics of fluids, more than half the articles are concerned with plasma physics, magnetohydrodynamics and fusion research and this is now the principal English language journal in this field.
7. *Planetary and Space Science.* Pergamon Press, Headington Hill Hall, Oxford. 1959 and onwards. (Monthly.)
   Includes occasional articles on natural plasmas.
8. *Plasma Physics, Accelerators, Thermonuclear Research. (Journal of Nuclear Energy, Pt C.)* Pergamon Press, Headington Hill Hall, Oxford. 1959 and onwards. (Bimonthly.)
   Original articles and research notes, mainly on plasma physics and fusion research. Includes translations of selected articles from *Atomnaya Energiya*.
9. *Zhurnal Tekhnicheskoi Fiziki.* Mendeleyevskaya lin. 1, Leningrad V-164. 1931 and onwards. (Monthly.)
   A large proportion of the articles are concerned with plasma physics and fusion research. Articles in Russian. Contents in Russian and English. (For cover-to-cover translation see Chapter 7.)
10. The following journals already mentioned in other chapters also contain papers on plasma physics and fusion research:
    American Physical Society Bulletin; Annals of Physics; Canadian Journal of Physics; Physical Review; Physical Review Letters; Proceedings of the Physical Society; Proceedings of the Royal Society; Zeitschrift für Naturforschung; Zhurnal Eksperimental'noi i Teoreticheskoi Fiziki.

SUBJECT INDEX

Abstracting services 117
Activation analysis 8
Advanced gas-cooled reactor 16
Astrophysics 210
Atomic bomb 125, 126
Atomic constants 151
Atomic Energy Act, 1946 (U.K.) 13
Atomic Energy Act, 1954 (U.S.A.) 42
Atomic Energy Authority Act, 1954 (U.K.) 13
Atomic Energy Control Act, 1946 (Canada) 24
Atomic physics 141

Bibliographies 117, 212
Bubble chambers 191

Cerenkov radiation 190
Chain reaction 4
Chemical processing 166
Cloud chambers 191
Cobalt-60 beam therapy units 25
Conference proceedings 126, 212
Controlled nuclear fusion 4, 210
Cosmic rays 148
Cover-to-cover translations 129
Criticality 166
Cyclotron 149

Decay period 7
Decay schemes 145
Dictionaries 123
Directories 120

Economic aspects of nuclear power 160
Effects of radiation 7, 8, 183, 184
Electrical discharges in gases 208
Electrodynamics 210
Electron 146
Elementary particles 10, 146
Encyclopaedias 122
Energy resources 3, 159
Environmental radioactivity 185, 186

Fall-out 185
Fast reactors 5, 162
Films on nuclear energy 19, 38, 39, 59, 99
Fission products 9
Fission reactions 145
Food preservation 183
Fossil fuels 3, 159
Fuel elements 166
Fuel processing 9, 166
Fundamental particles 146

Gas-cooled reactors 162
Gas discharges 208
Gaseous diffusion 8
Geiger counters 190
Glossaries 123
Graphite 168

Half-life 7
Handbooks 120
High energy physics 10, 147

Industrial radiography 180
Ionization chambers 190
Ionized gases 208
Ionizing radiation 7, 177
Isotope applications 178
Isotope catalogues 181
Isotope laboratory techniques 182
Isotope separation 8, 178
Isotope thickness gauges 180
Isotopes 6, 177

Linear accelerators 149
Liquid metals 168

MacMahon Act, 1946 13
Magnetohydrodynamics 209
Mass spectrometry 178
Mesons 147
Moderators 5

Neptunium 9
Neutrino 147
Neutron 146
Neutron detectors 191
Neutron diffraction 147
Neutron transport 147
Nuclear data 150
Nuclear data centres 151
Nuclear emulsions 191
Nuclear engineering 161
Nuclear fission 4, 145
Nuclear forces 144
Nuclear fusion 4, 203
Nuclear magnetism 144
Nuclear materials 167, 168
Nuclear models 144, 145
Nuclear moments 144
Nuclear physics 9, 142
Nuclear power costs 160
Nuclear power programmes 159
Nuclear power reactors 162
Nuclear properties 143
Nuclear propulsion 165
Nuclear reactions 145
Nuclear reactors 5, 161
Nuclear structure 9, 143

Particle accelerators 10, 149
Permissible dose 7, 188
Photographic emulsion detectors 191
Photon 146
Pitchblende 8
Plasma dynamics 209
Plasma physics 203, 208
Plutonium 5, 9, 168
Popular books 124, 142, 148, 150, 159, 160, 161, 170, 177, 178, 186, 211
Positron 147
Project Matterhorn 206
Project Sherwood 205
Protection *see* Radiation protection

Radiation chemistry 8, 183
Radiation detectors 190
Radiation dosimetry 189
Radiation injury 186
Radiation processing 183
Radiation protection 7, 187

Radiation sources 182
Radioactive standards 145
Radioactive wastes 169
Radioactivity 4, 7, 144
Radiobiology 184
Radiochemical laboratories 182
Radiochemistry 8, 182
Radiodiagnosis 180
Radioisotopes *see* Isotopes
Radiotherapy 181
Raw materials 8
Reactor catalogues 164
Reactor control 164
Reactor fuels 166
Reactor pressure vessels 163
Reactor safety 165
Reactor theory 163
Remote handling 7, 182
Report literature 19, 21, 26, 27, 29, 37, 38, 39, 40, 41, 48, 52, 58, 63, 67, 72, 75, 78, 82, 86, 87, 98, 103, 110
Rocket propulsion 165

Scattering 145
Scintillation counters 190
Shell model of the nucleus 144
Shielding 7, 166, 187
Ship propulsion 165
Strange particles 148
Subscription service, UKAEA 21

Thermal reactors 5, 161
Thermonuclear reactions *see* Nuclear fusion
Thorium 28, 167
Tracers 6, 177
Translations 51, 128
Transport of radioactive materials 189

Universal decimal classification 19, 49
Uranium 4, 5, 8, 9, 28, 167
Uranium-235 4, 5, 8, 168
U.S.—Euratom Joint Research Programme 102, 103

Waste disposal 9, 169

INDEX TO ORGANIZATIONS

AGIP Nucleare, Italy   65
AKK Atomic Power Group, Sweden   76
Aktiebolaget Atomenergi, Sweden   73
Allmanna Svenska Electriska, Sweden   76
All-Union Institute for Scientific and Technical Information, Moscow   48, 50
American Museum of Atomic Energy, Oak Ridge   35
American Nuclear Society   42
Ames Laboratory, Iowa   33
AMF Atomics (Canada) Ltd.   25
Argonne National Laboratory, Lamont   33
Armed Services Technical Information Agency   41
ASEA   *see* Allmanna Svenska Electriska
Aslib   130
Association Belge pour le Développement Pacifique de l'Énergie Atomique   78
Association Suisse pour l'Énergie Atomique   86
Association Technique pour la Production et l'Utilisation de l'Énergie Nucléaire   60
Associazione Nazionale di Ingegneria Nucleare   64
ASTIA   *see* Armed Services Technical Information Agency
Ateliers de Constructions Électriques de Charleroi   79
ATEN   *see* Association Technique pour la Production et l'Utilisation de l'Énergie Nucléaire
Atomic Energy Control Board, Canada   25, 26
Atomic Energy Establishment, Trombay   28
Atomic Energy Establishment, Winfrith   16
Atomic Energy of Canada, Ltd.   25
Atomic Energy Research Establishment, Harwell   14, 18, 20, 21
Atomic Energy Society of Japan   68
Atomic Fuel Corporation, Japan   67
Atomic Industrial Forum   24, 42
Atomic Weapons Research Establishment, Aldermaston   17, 20
Atomkraftkonsortiet Krangede, Sweden   76
Australian Atomic Energy Commission   26
Australian Institute of Nuclear Science and Engineering   27
Avco-Everett Research Laboratory, Massachusetts   206

Belgicatom   78
Belgonucléaire   78
Berkeley Nuclear Laboratories, Gloucestershire   22
Bettis Atomic Power Laboratory, Pittsburgh   33
Boeing Scientific Research Laboratories, Seattle   206
Boris Kidrich Institute of Nuclear Research, Belgrade   52
Brevatome   60
British Nuclear Energy Society   23
British Nuclear Forum   24
Brookhaven National Laboratory, Long Island   33

INDEX TO ORGANIZATIONS

Calder Hall 17
Calder Operations School 17
Canadian General Electric Co. 25
Capenhurst Works, Cheshire 17
Casaccia Centre for Nuclear Studies, Italy 63
Cavendish Laboratory, Cambridge 4
CENTO Institute of Nuclear Science, Tehran 29
Central Council for Nuclear Energy, Netherlands 79
Central Documentation Service, Saclay 58
Central Electricity Generating Board 22
Central Electricity Research Laboratories, Leatherhead 22
Central Institute of Physical Research, Csilleberc 52
Central Nuclear Measurements Bureau, Geel 102
Central Office of Information 13
Centre d'Étude de l'Énergie Nucléaire, Belgium 78
Centre d'Études Nucléaires de Cadarache 57
Centre d'Études Nucléaires de Fontenay-aux-Roses 57, 207
Centre d'Études Nucléaires de Grenoble 57
Centre d'Études Nucléaires de Saclay 57
Centre Européen de Traitement de l'Information Scientifique, Ispra 103
Centro Informazioni Studi Esperienze, Italy 64
CERN, Geneva 108
Chalk River, Canada 25
Chapelcross, Dumfriesshire 17
Comitato Nazionale per l'Energia Nucleare, Italy 61
Comitato Nazionale per le Richerche Nucleare, Italy 61
Commissariat à l'Énergie Atomique, Belgium 76
Commissariat à l'Énergie Atomique, France 55, 57
Commonwealth Index of Unpublished Scientific and Technical Translations 129
Computation Centre, Bologna 63
Conseil Européen pour la Recherche Nucléaire, Geneva *see* CERN
Culham Laboratory, Abingdon 16, 205
Czechoslovak Academy of Sciences 51
Czechoslovak Atomic Energy Commission 51

Danatom 88
Danish Atomic Energy Commission 87, 88
Daresbury Nuclear Physics Laboratory, Cheshire 22
Defence Documentation Center 41
Delegationen för Atomenergifrågor, Sweden 73
Department of Atomic Energy, India 28
Department of Scientific and Industrial Research 13
Depository libraries
  Australia 27
  Canada 26
  United Kingdom 21
  United States 40
Deutsche Atomkommission 69
Documentation Centre for Nuclear Energy, Netherlands 81
Dounreay Experimental Reactor Establishment, Caithness 17
Dragon Reactor Project, ENEA 107

École Polytechnique Fédérale, Zurich 85
Eidgenossisches Institut für Reaktorforschung, Würenlingen 85
Eldorado Mining and Refining Ltd., Canada 25
Electricité de France 55, 59
Ente Nazionale Idocarburi, Italy 64
Euratom 99, 206
   Common Research Centre 71, 81, 101, 102
   Transatom Service, Brussels 130
Eurochemic 106
European Atomic Energy Community 99, 206
European Atomic Forum 111
European Coal and Steel Community 99
European Common Market 99
European Company for Chemical Processing of Irradiated Fuels 106
European Neutron Data Compilation Centre, ENEA 107
European Nuclear Energy Agency 16, 104
European Organization for Nuclear Research, Geneva *see* CERN
European Translation Centre, Delft 130
European Transuranium Institute, Karlsruhe 102

Federal Commission for Nuclear Energy, Yugoslavia 52
Federal Commission for Nuclear Energy Affairs, Poland 52
Federal Institute for Reactor Research, Würenlingen 85
Fonds National de la Recherche Scientifique, Belgium 76
Foratom 111
Forum Italiano dell'Energia Nucleare 64
Foundation for Fundamental Research on Matter, Netherlands 79, 82, 207
Frascati National Laboratory, Italy 63

General Dynamics Corporation, San Diego 206
General Electric Research Laboratory, Schenectady 206
German Atomic Energy Commission 69
Gesellschaft für Kernenergieverwertung in Schiffbau und Schiffahrt m.b.H. 73
Gmelin Institute, Frankfurt 49, 71
Groupement Intersyndical de l'Industrie Nucléaire, France 60
Groupement Professionnel de l'Industrie Nucléaire, Belgium 78

Hahn-Meitner-Institut für Kernforschung, Berlin 72
Halden Reactor Project, ENEA 106
Hanford Laboratories, Richland 34
High Temperature Reactor Project, ENEA 107
Hungarian Academy of Sciences 52
Hydroelectric Power Commission, Ontario 25

Indian Atomic Energy Commission 27
Institut für Chemie, Mainz 72
Institut für Kernphysik, Heidelberg 72
Institut für Physik und Astrophysik, Munich 72, 207
Institut für Plasmaphysik, Garching 207

Institut für Plasmaphysik, Julich   207
Institute for Electro-physical Instruments, Leningrad   47, 206
Institute for Nuclear Physics Research, Amsterdam   82
Institute for Nuclear Study, Tokyo   68
Institute for Plasma Physics, Jutphaas   82, 207
Institute of Nuclear Physics, Novosibirsk   206
Institute of Nuclear Physics, Rez   51
Institute of Nuclear Research, Debrecen   52
Institute of Nuclear Research, Warsaw   52
Institute of Physics, Obninsk   47
Institute of Plasma Physics, Nagoya University   207
Institute of Plasma Physics, Prague   207
Institutt för Atomenergi, Norway   86, 87
International Atomic Energy Agency   19, 55, 97
International Book Centre   49, 50
International Federation for Documentation   19
Isotope School, Wantage   16
Ispra Centre for Nuclear Studies   63

Japan Atomic Energy Research Institute   65
Japan Atomic Industrial Forum   68
Japan Atomic Power Company   68
Japanese Atomic Energy Commission   65
Joint Committee on Atomic Energy, U.S. Congress   42
Joint Establishment for Nuclear Energy Research, Kjeller   79, 82, 86, 87
Joint Institute for Nuclear Research, Dubna   47
Josef Stefan Institute, Ljubljana   52
Julich Research Centre   71, 207

Karlsruhe Nuclear Research Centre   71
Kernforschungsanlage Jülich des Landes Nordrhein-Westfalen   71, 207
Kernforschungszentrum Karlsruhe   71
Kernreaktor Bau- und Betriebs-Gesellschaft m.b.H.   71
Kjeller Research Centre, Norway   87
Knolls Atomic Power Laboratory, Schenectady   34
Kurchatov Institute for Atomic Energy, Moscow   47, 206

Laboratoire de Recherches sur la Physique des Plasmas, Lausanne   85, 207
Laboratorio Gaz Ionizzati, Frascati   207
Laboratory for Mass Spectrometry, Amsterdam   82
Lawrence Radiation Laboratory, California   34, 206
Lebedev Institute, Moscow   47, 206
Library of Congress   50
Los Alamos Scientific Laboratory, New Mexico   34, 205
Lucas Heights Research Establishment, Sydney   26

Max-Planck-Gesellschaft zur Förderung der Wissenschaften   72
Micro Methods Ltd., East Ardsley   21
Microcard Editions Inc., West Salem   40

Ministry for the Construction of Electric Power Plants, U.S.S.R.   46
Ministry for Scientific Research, Germany   69
Ministry of Supply, U.K.   13, 14
Mol Research Centre, Belgium   78

National Aeronautics and Space Administration   41
National Bureau of Standards   41
National Institute for Research in Nuclear Science, U.K.   22
National Institute of Nuclear Physics, Italy   63
National Institute of Nuclear Science and Techniques, France   57
National Institute of Radiological Sciences, Japan   67
National Lending Library for Science and Technology   50, 129
National Reactor Testing Station, Idaho   34
National Research Council, Canada   26
Neratoom, N.V., Netherlands   83
Netherlands Institute for Documentation and Registration   81
Netherlands Nuclear Energy Committee   79
Netherlands Organization for Applied Scientific Research   83
Nuclear Computer Programme Library, ENEA   107
Nuclear Energy Trade Associations' Conference   23
Nuclear Research Foundation, University of Sydney   27

Oak Ridge Institute of Nuclear Studies   35
Oak Ridge National Laboratory, Tennessee   34
Office of Technical Services, U.S. Department of Commerce   40, 41
Organization for Economic Co-operation and Development   104

Pakistan Atomic Energy Commission   29
Pakistan Institute of Nuclear Science and Technology   29
Petten Research Centre, Netherlands   81
Physical Institute, Kiev   47
Physical Research Laboratory, Ahmedabad   28
Physico-Technical Institute, Kharkov   47, 206
Physico-Technical Institute, Leningrad   47
Plasma Physics Research Laboratory, Lausanne   85, 207
Polish Academy of Sciences   52
Princeton University   206

Radio Research Station, Slough   205
Radiochemical Centre, Amersham   16
Reactor Centrum Nederland   79, 81
Reactor Development Laboratories, Windscale   17
Reactor Fuel Element Laboratories, Springfields   17
Reactor Materials Laboratories, Culcheth   17
Reaktor AG, Switzerland   85
Reaktor-Sicherheitskommission, Germany   69
Risø Research Laboratory, Denmark   88, 207
Rudjer Boskovic Institute, Zagreb   52
Rutherford High Energy Laboratory, Chilton   22

Sandia Laboratory, Albuquerque   34
Savannah River Laboratory, Aiken   35
Schweizerische Gesellschaft von Fachleuten der Kerntechnik, Switzerland   86
Science Research Council, U.K.   22
Service Central de Documentation, Saclay   58
Siebersdorf Laboratory, Vienna   97
Societa per Azioni con Sede in Milano   65
Societa Italiana Meridionale Energia Atomica   65
Societa Richerche Impianti Nucleari   65
Société Belge Centre et Sud   59
Société Belge pour l'Industrie Nucléaire   78
Société d'Énergie Nucléaire Franco-Belge des Ardennes   59
Société Française pour la Gestion des Brevets d'Application Nucléaire   60
Société Nationale pour l'Encouragement de la Technique Atomique Industrielle, Switzerland   85
Special Libraries Association Translations Centre   129
Springfields Works, Lancashire   17
Statens Råd for Atomforksning, Sweden   73
Stitchting voor Fundamenteel Onderzoek der Materie, Netherlands   79, 82
Studiengesellschaft zur Förderung der Kernenergieverwertung in Schiffbau und Schiffahrt e.v.   73
Studsvik Research Establishment, Nyköping   75
Swedish Atomic Energy Board   73
Swedish Atomic Research Council   75
Swiss Federal Atomic Energy Commission   85
Swiss Nuclear Society   86

Takasaki Research Establishment, Japan   67
Tata Institute of Fundamental Research, Bombay   28
Technische Informationsbibliothek, Hanover   50, 72
Tokai Research Establishment, Japan   67
Tokyo University, Institute for Nuclear Study   68
Tot Keuring van Electrotechnische Materialien, Netherlands   83

United Kingdom Atomic Energy Authority   13
United States Atomic Energy Commission   20, 33
United States Naval Research Laboratory   206
U.S.S.R. Academy of Sciences   46, 47

VINITI   see All-Union Institute for Scientific and Technical Information

Wantage Research Laboratory, U.K.   16
Whiteshell Nuclear Research Establishment, Winnipeg   25
Wills Plasma Physics Laboratory, Sydney   207
Windscale Works, Cumberland   17

Zentralstelle für Atomkernenergie Dokumentation, Frankfurt   71

AUTHOR INDEX

ABBATT, J. D., AND OTHERS. *Protection against radiation* 197
ABRAGAM, A. *Principles of nuclear magnetism* 152
ADAIR, R. K., AND FOWLER, E. C. *Strange particles* 154
AKTIEBOLAGET ATOMENERGI.
  Aktiebolaget Atomenergi: the Atomic Energy Company of Sweden 91
  Annual report 91
ALEXANDER, P. *Atomic radiation and life* 197
ALFVEN, H., AND FÖLTHAMMAR, C. *Cosmical electrodynamics* 215
ALLARDICE, C., ED. *Atomic power: an appraisal* 171
ALLEN, C. W. *Astrophysical quantities* 216
ALLEN, J. S. *The neutrino* 154
ALLEN, W. D. *Neutron detection* 199
ALLIBONE, T. E. *The release and use of atomic energy* 135
ALLIS, W. P., ED. *Nuclear fusion* 216
ALLIS, W. P., AND OTHERS. *Waves in anisotropic plasmas* 215
ALMAN, M., ED. *Aslib directory* 132
AMERICAN CONFERENCE OF GOVERNMENTAL AND INDUSTRIAL HYGIENISTS. *Selected bibliography of radiation protection organisations* 198
AMERICAN INSTITUTE OF PHYSICS. *American Institute of Physics handbook* 133
AMERICAN STANDARDS ASSOCIATION. *Nuclear safety guide* 173
AMPHLETT, C. B. *Treatment and disposal of radioactive wastes* 175
ANDERSON, D. L. *The discovery of the electron* 154
Annuaire de l'Activité Nucléaire Française 89
ANTHONY, L. J. *Books on atomic energy* 131
ARMAND, L., AND OTHERS. *A target for Euratom* 112
ARMED SERVICES TECHNICAL INFORMATION AGENCY. *Plasma physics and magnetohydrodynamics: an ASTIA bibliography* 217
ASIMOV, I. *Inside the atom* 135
ASLIB. *British scientific and technical books* 134
ASSOCIATION SUISSE POUR L'ÉNERGIE ATOMIQUE. *Énergie atomique et protection contre les radiations en Suisse* 93
ASTON, F. W. *Mass spectra and isotopes* 192
ATOMIC ENERGY OF CANADA LTD. *List of publications* 31
ATOMIC ENERGY RESEARCH ESTABLISHMENT.
  A list of British books and periodicals on atomic energy 135
  A list of reports and published papers by A.E.R.E. staff 31
  A list of reports for sale and published papers by A.E.R.E. staff 31
  Atomic energy: government publications 31
  Harwell handbook 30
  List of A.E.R.E unclassified reports 31
  Selected abstracts of atomic energy project unclassified report literature in the field of radiation chemistry 196

BACON, G. E. *Neutron diffraction* 154
BACQ, Z. M., AND ALEXANDER, P. *Fundamentals of radiobiology* 196

BALDIN, A. M., AND OTHERS. *Kinematics of nuclear reactions* 153
BARNARD, G. P. *Modern mass spectrometry* 152
BARNES, D. E., AND TAYLOR, D. *Radiation hazards and protection* 197
BARNES, D. E., AND OTHERS. *Newnes concise encyclopaedia of nuclear energy* 133
BARTELS, H. *Die Forschungen auf dem Gebiet der Plasmaphysik in der Bundesrepublik Deutschland* 214
BEHRENS, C. F., ED. *Atomic medicine* 197
BEIERWALTES, W. H., AND OTHERS. *Clinical use of radioisotopes* 194
BELL, C. G., AND HAYES, F. N., EDS. *Liquid scintillation counting* 199
BENE, G. J., AND OTHERS. *Nuclear physics and atomic energy* 134
BERSHADER, D., ED. *Plasma hydromagnetics* 216
BETHE, H. A., AND MORRISON, P. *Elementary nuclear theory* 152
BEYER, R. T., ED. *Foundations of nuclear physics* 151
BIRKS, J. B. *Scintillation counters* 199
BIRKS, J. B., ED. *Proceedings of the symposium on nuclear instruments* 199
BISHOP, A. S. *Project Sherwood* 216
BLACKWOOD, O. H., AND OTHERS. *An outline of atomic physics* 151
BLATT, J. M., AND WEISSKOPF, V. F. *Theoretical nuclear physics* 151
BLATZ, H., ED. *Radiation hygiene handbook* 197
BLIN-STOYLE, R. G. *Theories of nuclear moments* 152
BOCKASTON, K., AND OTHERS. *Controlled thermonuclear fusion research* 216
BOEING SCIENTIFIC RESEARCH LABORATORIES. PLASMA PHYSICS LABORATORY. *Progress review* 214
BOLT, R. O., AND CARROLL, J. G. *Radiation effects on organic materials* 196
BORN, M. *Atomic physics* 151
BOST, W. E. *Radiation protection standards: a literature search* 198
BOWEN, H. J. M., AND GIBBONS, D. *Radioactivation analysis* 193
BOWEN, J. H., AND MASTERS, E. F. O. *Nuclear reactor control and instrumentation* 172
BRADLEY, J. E. S., TRANS. *Physics of nuclear fission* 153
BRAESTRUP, C. B., AND WYCKOFF, H. O. *Radiation protection* 197
BRANDSTATTER, J. J. *An introduction to waves, rays and radiation in plasma media* 215
BREE, R. *Regional co-operation in the field of scientific and technical information within the European Community* 112
BRITISH NATIONAL BIBLIOGRAPHY. *Cumulated subject catalogue* 134
BRITISH STANDARDS INSTITUTION. *Glossary of terms used in nuclear science* 134
BRODA, E. *Radioactive isotopes in biochemistry* 194
BROMLEY, D. A., AND VOGT, E. W., EDS. *Proceedings of the International Conference on Nuclear Structure* 152
BROWN, S. C. *Basic data of plasma physics* 215
BROWNING, E. *Harmful effects of ionizing radiations* 197
BURCHAM, W. E. *Nuclear physics* 152
BUSH, H. D. *Atomic and nuclear physics* 151
BUSSARD, R. W., COMP. *Nuclear rocket propulsion: a selected bibliography* 173
BUSSARD, R. W., AND DE LAUER, R. D. *Nuclear rocket propulsion* 173
BUTLER, S. T., AND HITTMAIR, O. H. *Nuclear stripping reactions* 153

CAMERON, J. F.
  Backscattering gauges 193
  Transmission gauges 193
CARO, D. E., AND OTHERS. *Modern physics* 135

CATTELL, J., ED. *American men of science* 133
CENTRAL ELECTRICITY GENERATING BOARD. *Berkeley nuclear laboratories* 31
CENTRAL OFFICE OF INFORMATION. *Nuclear energy in Britain* 30
CENTRE D'ÉTUDES DE L'ÉNERGIE NUCLEAIRE.
  Mol 92
  Rapport sur l'exercise 1961 92
  The development of nuclear energy in Belgium 92
CENTRO INFORMAZIONI STUDI ESPERIENZE.
  CISE laboratories 90
  List of CISE's publications 90
CHADWICK, SIR J. *Radioactivity and radioactive substances* 153
CHANDRASEKHAR, S. *Hydrodynamic and hydromagnetic stability* 215
CHARLESBY, A. *Atomic radiation and polymers* 196
CHARLESBY, A., ED. *Radiation sources* 195
CHASTAIN, J. W., ED. *U.S. research reactor operation and use* 135
CLASON, W. E., ED. *Elsevier's dictionary of nuclear science and technology* 134
CLAUS, W. D., ED. *Radiation biology and medicine* 196
CLAUSER, F. H. *Symposium on plasma dynamics* 215
CLEGG, J. W., AND FOLEY, D. D., EDS. *Uranium ore processing* 174
CLEMENTEL, E., AND VILLI, C., EDS. *Proceedings of the Conference on Direct Reactions and Nuclear Reaction Mechanisms* 153
COBINE, J. D. *Gaseous conductors* 214
COFFINBERRY, A. S., AND MINER, W. N., EDS. *The metal plutonium* 174
COLLINS, J. C., ED. *Radioactive wastes* 174
COMAR, C. L. *Radioisotopes in biology and agriculture* 194
COMITATO NAZIONALE PER L'ENERGIA NUCLEARI.
  Il Centro della Casaccia 90
  Rapporti tecnici: catalogo 90
  Rapporto di attivita 89
COMITATO NAZIONALE PER L'ENERGIA NUCLEARI. LABORATORI NAZIONALI DI FRASCATI.
  List of reports and abstracts, 1953–1961 90
  Activity at the National Laboratories, Frascati 90
COMITATO NAZIONALE PER LE RICHERCHE NUCLEARI.
  Activities report 90
  A five year plan for the development of nuclear research in Italy 89
  CNRN 1952–1959 89
  The centre for Nuclear Studies, Ispra 90
  The National Synchrotron Laboratory, Frascati 90
COMMISSARIAT A L'ÉNERGIE ATOMIQUE.
  Cinemathèque 89
  Developments and programmes 89
  La documentation scientifique et technique au CEA 89
  Liste récapitulative des rapports CEA publiés 1948–1961 89
  Rapport annuel 88
  The Cadarache nuclear research centre 88
  The Fontenay-aux-Roses nuclear research centre 88
  The French Atomic Energy Commission 1945–1960 88
  The Grenoble nuclear research centre 88
  The Saclay nuclear research centre 88
COMMISSARIAT A L'ÉNERGIE ATOMIQUE. CENTRE D'ÉTUDES NUCLEAIRES DE GRENOBLE.
  Publications et notes scientifiques 89
CONDON, E. U., AND ODISHAW, H., EDS. *Handbook of physics* 133

COOPER, A. R. *Chemical processing in the atomic energy industry* 174
COWLING, T. G. *Magnetohydrodynamics* 215
CRANSHAW, T. E. *Cosmic rays* 155
CRONKITE, E. P., AND BOND, V. P. *Radiation injury in man* 197
CROUCH, H. F. *Nuclear ship propulsion* 173
CURRAN, S. C. *Luminescence and the scintillation counter* 199
CURRY, D., AND NEWMAN, B. R. *The challenge of fusion* 216
CURTISS, L. F. *Introduction to neutron physics* 154
CUTHBERT, F. L. *Thorium production technology* 174

DALITZ, R. H. *Strange particles and strong interactions* 154
DANISH ACADEMY OF TECHNICAL SCIENCES. *Danatom annual report* 93
DANISH ATOMIC ENERGY COMMISSION. *Report on activities* 93
DAUDEL, P. *Radioactive tracers in chemistry and industry* 193
DAVISON, B., AND SYKES, J. B. *Neutron transport theory* 154
DEARNALEY, G., AND NORTHROP, D. C. *Semiconductor counters for nuclear radiations* 199
DEFENSE DOCUMENTATION CENTER. *Plasma physics and magnetohydrodynamics: a report bibliography* 217
DELARIO, A. J. *Roentgen, radium and radioisotope therapy* 194
DELCROIX, J. L. *Introduction to the theory of ionized gases* 215
DENISSE, J. F., AND DELCROIX, J. L. *Plasma waves* 215
DEPARTMENT OF SCIENTIFIC AND INDUSTRIAL RESEARCH.
  Radio research: report of the Radio Research Board 213
  Scientific and technical information in the Soviet Union 54
DEPARTMENT OF SCIENTIFIC AND INDUSTRIAL RESEARCH. LENDING LIBRARY UNIT. *Titles of periodicals from the U.S.S.R.* 53
DE SHALIT, A., AND TALMI, I. *Nuclear shell theory* 152
DESROSIER, N. W., AND ROSENSTOCK, H. M. *Radiation technology in food, agriculture and biology* 195
DICK, W. E. *Atomic energy in agriculture* 194
DIETRICH, J. R., AND ZINN, W. H. *Solid fuel reactors* 135
Directory of British scientists 133
DRUMMOND, J. E. *Plasma physics* 215
DUCKWORTH, H. E. *Mass spectroscopy* 152
DUCKWORTH, H. E., ED. *Proceedings of the International Conference on Nuclidic Masses* 152
DUNCAN, J. F., AND COOK, G. B. *Modern radiochemical practice* 195
DUNGEY, J. W. *Cosmic electrodynamics* 215
DZHELEPOV, B. S., AND PEKER, L. K. *Decay schemes of radioactive nuclei* 153

EAVES, G. *Principles of radiation protection* 197
EISBERG, R. M. *Fundamentals of modern physics* 151
EISENBUD, L., AND WIGNER, E. P. *Nuclear structure* 152
EISENBUD, M. *Environmental radioactivity* 197
ÉLECTRICITE DE FRANCE.
  La central nucléaire de Chinon 89
  Rapport d'activité 89
ELTON, L. R. B.
  Introductory nuclear theory 151
  Nuclear sizes 152

EL-WABIL, M. M. *Nuclear power engineering* 171
ENDT, P. M., AND OTHERS. *Nuclear reactions* 153
ETHERINGTON, H., ED. *Nuclear engineering handbook* 133
ÉTIEVANT, C. *L'énergie thermonucléaire* 216
ETTER, L. E. *Glossary of words and phrases used in radiology and nuclear medicine* 194
EURATOM.
   Euratom: the European Atomic Energy Community 112
   Euratom keyword thesaurus 113
   General report on the activities of the Community 112
   List of the scientific and technical reports published 113
   Nuclear installations in the countries of the European Atomic Energy Community 172
   Treaty establishing the European Atomic Energy Community 112
EURATOM SUPPLY AGENCY. *State of the nuclear fuel market in the Community* 171
EURATOM—CEA. GROUP DE RECHERCHES SUR LA FUSION.
   Liste des publications 214
   Rapport d'activité du Group de Recherches 214
EURATOM—KFA. INSTITUT FÜR PLASMAPHYSIK. *Arbeitsberichte der theoretischen und experimentellen Gruppen* 214
EUROCHEMIC. *First activity report* 113
EUROPEAN ATOMIC ENERGY COMMUNITY *see* EURATOM
EUROPEAN NUCLEAR ENERGY AGENCY.
   Catalogue of courses on nuclear energy 114
   Radiation protection norms 198
   Report on the activities of the Agency 113
   Statute of the Agency 113
   The OEEC Nuclear Energy Agency: structure and functions 113
EUROPEAN ORGANISATION FOR NUCLEAR RESEARCH.
   Annual report 114
   Catalogue of periodicals in the CERN library 114
   Guide to the CERN library 114
   Index of scientific publications 114
EVANS, R. D. *The atomic nucleus* 151
EXTERMANN, R. C., ED. *Radioisotopes in scientific research* 193

FAIRES, R. A., AND PARKS, B. H. *Radioisotope laboratory techniques* 195
FARMER, F. R. *The packaging, transport and related handling of radioactive materials* 198
FEARNSIDE, K., AND OTHERS. *Applied atomic energy* 192
FEENBERG, E. *Shell theory of the nucleus* 152
FERRARO, U. C. A., AND PLUMPTON, C. *An introduction to magnetofluidmechanics* 215
FERRETTI, B., ED. *Proceedings of the 1958 Annual International Conference on High Energy Physics, CERN* 154
FIELDS, T., AND SEED, L., EDS. *Clinical use of radioisotopes* 194
FLAGG, J. F., ED. *Chemical processing of reactor fuels* 174
FLUGGE, S., ED. *Handbuch der Physik* 133
FOOD AND AGRICULTURE ORGANISATION. *Radioactive materials in food and agriculture* 197
FOZARD, B. *Instrumentation and control of nuclear reactors* 172
FRANCIS, G. *Ionization phenomena in gases* 215
FRANCIS, G. E., AND OTHERS. *Isotopic tracers* 194

FRANCIS, L. S., COMP. *Magnetic mirror machines* 217
FREMLIN, J. H. *Applications of nuclear physics* 199
FRIEDLANDER, G., AND OTHERS. *Nuclear and radiochemistry* 195
FRISCH, O. R. *Atomic physics today* 151
FRISCH, O. R., ED. *The nuclear handbook* 133
FROST, B. R. T., AND WALDRON, M. B. *Nuclear reactor materials* 174
FUTTERMANN, W. I., ED. *Propagation and instabilities in plasmas* 217

GALBRAITH, W. *Extensive air showers* 155
GARTENHAUS, S. *Elements of plasma physics* 215
GIBBS, R. C., AND WAY, K., EDS. *A directory to nuclear data tabulations* 155
GILLESPIE, A. B. *Signal, noise and resolution in nuclear counter amplifiers* 199
GINZBURG, V. L. *Propagation of electromagnetic waves in plasma* 215
GITTUS, J. H. *Uranium* 174
GLASCOCK, R. *Labelled atoms* 192
GLASGOW. ROYAL COLLEGE OF SCIENCE AND TECHNOLOGY. *Nuclear reactor containment buildings and pressure vessels* 172
GLASSNER, A. *Introduction to nuclear science* 135
GLASSTONE, S. *Sourcebook on atomic energy* 135
GLASSTONE, S., ED. *The effects of nuclear weapons* 197
GLASSTONE, S., AND EDLUND, M. C. *Elements of nuclear reactor theory* 172
GLASSTONE, S., AND LOVBERG, R. H. *Controlled thermonuclear reactions* 216
GLASSTONE, S., AND SESONSKE, A. *Nuclear reactor engineering* 171
GLUECKAUF, E., ED. *Atomic energy waste* 175
GOLDRING, M.
  Economics of atomic energy 171
  Replanning Britain's nuclear power programme 170
GOROKHOFF, B. I. *Providing U.S. scientists with Soviet scientific information* 138
GOVE, N. B. *Information centres in nuclear physics* 156
GOWING, M. *Britain and atomic energy* 136
GRAINGER, L. *Uranium and thorium* 174
GREAT BRITAIN.
  FOREIGN OFFICE. *Decision of the Council of OEEC establishing a European Nuclear Energy Agency* 113
  LORD PRESIDENT OF THE COUNCIL AND MINISTER OF FUEL AND POWER. *A programme of nuclear power* 170
  MINISTRY OF HOUSING AND LOCAL GOVERNMENT. *The control of radioactive wastes* 175
  MINISTRY OF LABOUR. *Code of practice for the protection of persons exposed to ionizing radiations* 198
  MINISTRY OF POWER. *The nuclear power programme* 170
  MINISTRY OF POWER. *The second nuclear power programme* 170
  MINISTRY OF SUPPLY. *Harwell, the British Atomic Energy Establishment* 30
  MINISTRY OF SUPPLY. *Radioisotopes techniques* 192
  SECRETARY OF STATE FOR SCOTLAND AND MINISTER OF POWER. *Capital investment in the coal, gas and electricity industries* 170
  STATUTES. *Radioactive Substances Act, 1960* 175
GREEN, A. E. S. *Nuclear physics* 151
GREEN, T. S. *Thermonuclear power* 216
GRIFFITH, T. C., AND POWER, E. A., EDS. *Nuclear forces and the few-nucleon problem* 152
GROUPEMENT PROFESSIONNEL DE L'INDUSTRIE NUCLEAIRE. *Repertory of the nuclear institutions and industry in Belgium* 92

HAHN, O., AND STRASSMANN, F. *Über den Nachweis und das Verhalten der bei der Bestrahlung des Urans mittels Neutronen entstehenden Erdalkalimetalle* 91
HALL, J. *The story of the construction of Berkeley Nuclear Power Station* 171
HALLIDAY, D. *Introductory nuclear physics* 152
HALNAN, K. E. *Atomic energy in medicine* 194
HAMILTON, J. *The theory of elementary particles* 154
HAMMOND, R. *British nuclear power stations* 171
HANDLOSER, J. S. *Health physics instrumentation* 199
HANFORD ATOMIC PRODUCTS OPERATION. *Review of power and heat reactor designs* 173
HANSON, N. R. *The concept of the positron* 154
HARNWELL, G. P., AND STEPHENS, W. E. *Atomic physics* 151
HARRER, J. M. *Nuclear reactor control engineering* 172
HARRINGTON, C. D., AND RUEHLE, A. E., EDS. *Uranium production technology* 174
HARRIS, R. J. C., ED. *The initial effects of ionizing radiation on cells* 196
HARRISON, J. R. *Nuclear reactor shielding* 173
HART, H., AND OTHERS. *Radioaktive Isotope in der Betriebsmesstechnik* 193
HARVEY, B. G. *Introduction to nuclear physics and chemistry* 152
HARVEY, J. F. *Pressure vessel design* 172
HASLETT, A. W., ED.
  Who's who in atoms 133
  World nuclear directory 132
HASTED, J. B. *Physics of atomic collisions* 215
HAWKINS, R. R. *Scientific, medical and technical books* 135
HEVESY, G. *Radioactive indicators* 193
HEVESY, G., AND PANETH, F. A. *A manual of radioactivity* 153
HEWLETT, R. G., AND ANDERSON, O. E. *The new world, 1939—1946* 136
HIGATSBERGER, M. J., AND VIEHBÖCK, F. P., EDS. *Electromagnetic separation of radioactive isotopes* 192
High energy nuclear physics: proceedings of the annual Rochester Conference 154
HILL, J. F. *Textbook of reactor physics* 172
HILL, R. D. *Tracking down particles* 155
HINE, G. J., AND BROWNELL, G. L., EDS. *Radiation dosimetry* 198
HINTENBERGER, H., ED. *Nuclear masses and their determination* 152
HODGMAN, C. D., AND OTHERS. *Handbook of chemistry and physics* 133
HODGSON, P. E. *The optical model of elastic scattering* 153
HOGERTON, J. F., AND OTHERS. *The atomic energy deskbook* 133
HOISINGTON, D. B. *Nucleonics fundamentals* 135
HOLDEN, A. N. *Physical metallurgy of uranium* 174
HOLLAENDER, A., ED.
  Radiation biology 196
  Radiation protection and recovery 197
HOOPER, J. E., AND SCHARFF, M. *The cosmic radiation* 155
HUGHES, D. J.
  Neutron cross sections 172
  Neutron optics 154
HUGHES, D. J., AND SCHWARTZ, R. B. *Neutron cross sections* 172
HYDE, E. K., *A review of nuclear fission* 153
HYDE, E. K., AND OTHERS. *Nuclear properties of the heavy elements* 153

ILLINOIS INSTITUTE OF TECHNOLOGY. *Literature survey on world isotope and radiation technology* 193
INSTITUT FÜR PLASMAPHYSIK, GARCHING. *Jahresbericht des Instituts* 214

INSTITUTE NATIONAL DES SCIENCES ET TECHNIQUES NUCLEAIRES. *Génie atomique* 133
INSTITUTT FOR ATOMENERGI.
  Survey of activities 1948—1960 93
  The Institutt for Atomenergi, Kjeller and Halden 93
INTERNATIONAL ATOMIC ENERGY AGENCY.
  Annual report of the Board of Governors 111
  Annual report to the Economic and Social Council of the U.N. 111
  Basic safety standards for radiation protection 198
  Directory of nuclear reactors 172
  Disposal of radioactive wastes 175
  Films on the peaceful uses of atomic energy 112
  I.A.E.A. research contracts 112
  Industrial uses of large radiation sources 195
  International directory of radioisotopes 195
  Large radiation sources in industry 195
  List of periodicals in the field of nuclear energy 112
  Metrology of radionuclides 153
  Nuclear propulsion 173
  Nuclear reactors 171
  Nuclear ship propulsion 173
  Operating experience with power reactors 171
  Proceedings of the Conference on Plasma Physics and Controlled Nuclear Fusion Research, Salzburg, 1961 216
  Production and use of short-lived radioisotopes from reactors 192
  Publications in the nuclear sciences 112
  Radioactive waste disposal into the sea 175
  Radioisotope applications in industry 193
  Radioisotopes in the physical sciences and industry 193
  Regulations for the safe transport of radioactive materials 198
  Research on controlled thermonuclear fusion 217
  Safe handling of radioisotopes 198
  Safe operation of critical assemblies and research reactors 173
  Treatment and storage of high-level radioactive wastes 175
INTERNATIONAL COMMISSION ON RADIOLOGICAL PROTECTION. *Radiation protection: recommendations* 198
INTERNATIONAL CONFERENCE ON IONIZATION PHENOMENA IN GASES, 4TH, UPPSALA, 1959. *Proceedings* 216
INTERNATIONAL CONFERENCE ON IONIZATION PHENOMENA IN GASES, 5TH, MUNICH, 1961. *Proceedings* 216
INTERNATIONAL CONFERENCE ON IONIZATION PHENOMENA IN GASES, 6TH, PARIS, 1963. *Proceedings* 216
INTERNATIONAL CONFERENCE ON THE PEACEFUL USES OF ATOMIC ENERGY, GENEVA, 1955. *Proceedings* 135
INTERNATIONAL CONFERENCE ON THE PEACEFUL USES OF ATOMIC ENERGY, GENEVA, 1958. *Proceedings* 136
INTERNATIONAL FEDERATION FOR DOCUMENTATION. *Universal decimal classification: special subject edition for nuclear science* 30
INTERNATIONAL LABOUR OFFICE. *Manual of industrial radiation protection* 198
ISBIN, H. S. *Introductory nuclear reactor theory* 172

JACKSON, C. B., ED. *Liquid metals handbook: sodium—NaK supplement* 174
JACKSON, J. D. *The physics of elementary particles* 154

Janossy, L. *Cosmic rays* 155
Japan Atomic Energy Commission.
  Atomic energy in Japan 90
  Japan Atomic Energy Research Institute: a brief guide 90
  Long range programme on development and utilisation of atomic energy 90
Japan Atomic Industrial Forum.
  Atomic energy yearbook 91
  JAIF's directory to governmental agencies, associations, etc. 91
Jay, K. E. B.
  Atomic energy research at Harwell 30
  Britain's atomic factories 30
  Calder Hall 30
  Nuclear power today and tomorrow 171
Jefferson, S. *Radioisotopes: a new tool for industry* 192
Jefferson, S., ed.
  Handbook of the atomic energy industry 133
  Massive radiation techniques 195
Jelley, J. V. *Cerenkov radiation* 199
Johnston, J. E., and others. *Radioisotope Conference, 1954* 192
Joint Committee on Atomic Energy. *Review of the international atomic policies and programs of the U.S.* 45
Jones, G. O., and others. *Atoms and the universe* 135
Jones, P. B. *Optical model in nuclear and particle physics* 153
Jukes, J. D. *Man-made sun* 216

Kaindl, K., and Linser, H. *Radiation in agricultural research and practice* 194
Kamen, M. D. *Isotopic tracers in biology* 194
Kaplan, I. *Nuclear physics* 151
Karetnikov, D. V., and others. *Linear ion accelerators* 155
Kaufmann, A. R., ed. *Nuclear reactor fuel elements* 173
Kernforschungszentrum Karlsruhe. *Liste der wissenschaftlichen Veröffentlichungen des Kernforschungszentrum Karlsruhe* 91
King, E. R., and Mitchell, T. G. *A manual for nuclear medicine* 194
Kircher, J. F., and Bowman, R. E. *Effects of radiation on materials and components* 196
Kistermaker, J., and others, eds. *Proceedings of the International Symposium on Isotope Separation, Amsterdam, 1957* 192
Klemperer, O., *Electron physics* 154
Knox, F. B., and Sabel, C. S. *Oscillating electromagnetic field interaction with, and containment of, plasmas* 217
Koch, J., and others. *Electromagnetic isotope separators and applications of electromagnetically enriched isotopes* 192
Kohl, J., and others. *Radioisotope applications engineering* 193
Komarovski, A. N. *Shielding materials for nuclear reactors* 173
Kopelmann, B., ed. *Materials for nuclear reactors* 174
Kopfermann, H. *Nuclear moments* 152
Kowarski, L. *An account of the origin and beginnings of CERN* 114
Kramer, A. W.
  Boiling water reactors 135
  Nuclear propulsion for merchant ships 173
Kramish, A. *Atomic energy in the Soviet Union* 53
Kunz, W., and Schintlmeister, J. P. *Nuclear tables* 156
Kuzin, A. M. *The application of radioisotopes in biology* 193

LADD, M. F. C., AND LEE, W. H. *Practical radiochemistry* 195
LAMBIE, D. A. *Techniques for the use of radioisotopes in analysis* 193
LANDSHOFF, R. K. M., ED. *The plasma in a magnetic field* 216
LANE, J. A., AND OTHERS, EDS. *Fluid fuel reactors* 135
LANGMUIR, I. *The collected works of Irving Langmuir* 214
LANSDELL, N. *The atom and the energy revolution* 170
LAPP, R. E., AND ANDREWS, H. L. *Nuclear radiation physics* 192
LAURENCE, W. L. *Dawn over zero* 136
LEA, D. E. *Actions of radiations on living cells* 196
LEONTOVICH, M. A., ED.
    Plasma physics and the problem of controlled thermonuclear reactions 216
    Problems in plasma theory 215
LEPRINCE-RINGUET, L. *Cosmic rays* 155
LIND, S. C., AND OTHERS. *Radiation chemistry of gases* 196
LINHART, J. G. *Plasma physics* 215
LIPKIN, H. J., ED. *Proceedings of the Rehovoth Conference on Nuclear Structure* 152
LITTLEFIELD, T. A., AND THORNLEY, N. *Atomic and nuclear physics* 151
LITTLER, D. J., AND RAFFLE, J. F. *An introduction to reactor physics* 172
LIVERHART, S. E. *Elementary introduction to nuclear reactor physics* 172
LIVINGOOD, J. J. *Principles of cyclic particle accelerators* 155
LIVINGSTON, M. S., AND BLEWETT, J. P. *Particle accelerators* 155
LLEWELLYN-JONES, F. *Ionization and breakdown in gases* 215
LOCK, W. O. *High energy nuclear physics* 154
LOEB, L. B. *Basic processes of gaseous electronics* 215
LOFTNESS, R. L. *Nuclear power plants* 171
LONDON, H., ED. *Separation of isotopes* 192
LONG, G., AND OTHERS. *The new power* 171
LONGMIRE, C. L. *Elementary plasma physics* 215
LOS ALAMOS SCIENTIFIC LABORATORY. *Semiannual status report of the LASL controlled thermonuclear research program* 213
LOUTIT, J. F. *Irradiation of mice and men* 197
LYON, R. N., ED. *Liquid metals handbook* 174

McCULLOUGH, C. R., ED. *Safety aspects of nuclear reactors* 173
McDOWELL, C. A. *Mass spectrometry* 152
McINTOSH, A. B., AND HEAL, T. J., EDS. *Materials for nuclear engineers* 174
McLEAN, A. S., AND OTHERS. *Radiation levels* 197
MALMBERG, C. *Abstract bibliography on linear accelerators* 155
MANN, W. B. *The cyclotron* 155
MANSFIELD, W. K. *Elementary nuclear physics* 152
MARSHAK, R. E. *Meson physics* 154
MARSHAK, R. E., AND SUDERSHAN, E. C. G. *Introduction to elementary particle physics* 153
MARTIN, F. S., AND MILES, G. L. *Chemical processing of nuclear fuels* 173
MASSEY, H. S. W., AND BURHOP, E. H. S. *Electronic and ionic impact phenomena* 215
MATHER, K. B., AND SWAN, P. *Nuclear scattering* 153
MEDICAL RESEARCH COUNCIL.
    Introductory manual on the control of health hazards from radioactive materials 197
    The hazards to man of nuclear and allied radiations 196
MELIK-SHAKHNAZAROV, A. S. *Technical information in the U.S.S.R.* 54
MEYER, L. *Atomic energy in industry* 135

Meyer, M. G., and Jensen, J. H. D. *Elementary theory of nuclear shell structure* 152
Michels, W. C., and others, eds. *International dictionary of physics and electronics* 134
Millikan, R. A.
  Electrons, protons, photons, neutrons, mesotrons and cosmic rays 154
  The electron 154
Mills, M. M., and others, eds. *Modern nuclear technology* 135
Mitchner, M. E. D. *Radiation and waves in plasmas* 216
Modelski, G. A. *Atomic energy in the Communist Bloc* 53
Montgomery, D. J. *Cosmic ray physics* 155
Mullenbach, P. *Civilian nuclear power* 170
Murphy, G. *Elements of nuclear engineering* 171
Murray, R. L.
  Introduction to nuclear engineering 171
  Nuclear reactor physics 172

Nagoya University. Institute of Plasma Physics. *Annual review* 214
Nakicenovic, S. *Nuclear energy in Yugoslavia* 54
Napolitano, L. G., and Contursi, G. *Magnetofluiddynamics: current papers and abstracts* 217
National Academy of Sciences. National Research Council. *The biological effects of atomic radiation* 196
National Bureau of Standards.
  A manual of radioactivity procedures 197
  Publications of the N.B.S. 45
National Federation of Science Abstracting and Indexing Services. *A guide to the world's abstracting and indexing services* 132
National Lending Library. *List of irregular serials received from the U.S.S.R. and Bulgaria* 54
National Research Council.
  A glossary of terms in nuclear science and technology 134
  The biological effects of atomic radiation 196
National Science Foundation.
  A guide to the scientific and technical literature of Eastern Europe 54
  Directory of selected scientific institutions in the U.S.S.R. 132
  List of Russian scientific journals available in English 137
  Specialised science information services in the U.S. 132
Nemirovskii, P. E.
  Contemporary models of the atomic nucleus 152
  Nuclear models 152
Netherlands-Norwegian Joint Establishment for Nuclear Energy Research. *Annual report* 93
Netherlands Nuclear Energy Committee. *A survey of the nuclear energy organisation in the Netherlands* 92
Nightingale, R. E., ed. *Nuclear graphite* 174
Nishijima, K. *Fundamental particles* 154

Oak Ridge National Laboratory.
  Catalog and price list of stable isotopes 195
  Radioisotopes, special materials and services 195
  Thermonuclear Division semiannual progress report 213

ORGANISATION FOR ECONOMIC CO-OPERATION AND DEVELOPMENT. HALDEN REACTOR
  PROJECT.
  Annual report 113
  Halden heavy boiling water reactor 113
ORGANISATION FOR ECONOMIC CO-OPERATION AND DEVELOPMENT. HIGH TEMPERATURE
  REACTOR PROJECT. *Annual report* 113
ORGANISATION FOR EUROPEAN ECONOMIC CO-OPERATION. *The industrial challenge
  of nuclear energy* 113
OUGHTERSON, A. W., AND WARREN, S. *Medical effects of the atomic bomb in
  Japan* 196
OVERMAN, R. T., AND CLARK, H. M. *Radioisotope techniques* 195

PALMER, R. G., AND PLATT, A. *Fast reactors* 171
PALUMBO, D. *Das Euratom Program zur Fusionsforschung* 214
PATERSON, R. *The treatment of malignant disease by radiotherapy* 194
PATTON, F. S., AND OTHERS. *Enriched uranium processing* 174
PEARSON, F. J., ED. *Nuclear power technology* 171
PETERSON, S., AND WYMER, R. G. *Chemistry in nuclear technology* 174
PICKARD, J. F., ED. *Nuclear power reactors* 171
PICKARD, J. F., AND OTHERS, EDS. *Power reactor technology* 171
PIRIE, A., ED. *Fall-out* 197
POST, R. F.
  Controlled fusion research 216
  The fourth state of matter 216
POULTER, D. R., ED. *Design of gas-cooled, graphite-moderated reactors* 171
POWELL, C. F., AND OTHERS. *The study of elementary particles by the photographic
  method* 199
PRENTKI, J., ED. *Proceedings of the 1962 International Conference on High-Energy
  Physics at CERN* 154
PRESTON, M. A. *Physics of the nucleus* 152
PRICE, B. T., AND OTHERS. *Radiation shielding* 173
PRICE, W. J. *Nuclear radiation detection* 199
PRINCETON UNIVERSITY. PLASMA PHYSICS LABORATORY.
  Annual report 213
  Cumulative listing of Plasma Physics Laboratory reports and publications 214
PURDOM, C. E. *Genetic effects of radiation* 197
PUTMAN, J. L. *Isotopes* 192
PUTNAM, P. C. *Energy in the future* 170

QUIMBY, E. N. *Safe handling of radioactive isotopes in medical practice* 197
QUIMBY, E. N., AND OTHERS. *Radioactive isotopes in medicine and biology* 194

RAMAKRISHNAN, A. *Elementary particles and cosmic rays* 155
RAMSAY, N. F. *Nuclear magnetic moments* 152
RATCLIFFE, J. A. *The magneto-ionic theory and its application to the ionosphere* 215
RATNER, B. S. *Accelerators of charged particles* 155
REACTOR CENTRUM NEDERLAND.
  Summary of activities 92
  The research centre of Reactor Centrum Nederland 92
  Reactor handbook 171
RITSON, D. M. *Techniques of high energy physics* 155
ROBERTS, J. E. *Codes of practice for the use of ionizing radiations* 198

AUTHOR INDEX

ROBERTSON, H. J. *Mass spectrometry* 152
ROCHLIN, R. S., AND SCHULTZ, W. W. *Radioisotopes for industry* 193
ROCHMANN, A. *Introduction to nuclear power costs* 171
ROCKLEY, J. C. *An introduction to industrial radiology* 193
ROSE, D. J., AND CLARK, M. *Plasmas and controlled fusion* 215
ROSENFELD, L. *Nuclear forces* 152
ROSSI, B. *High energy particles* 154
ROTBLAT, J., AND OTHERS.
  The acceleration of particles to high energies 155
  The uses and effects of nuclear energy 192
RUSSEL, C. R. *Reactor safeguards* 173
RUTHERFORD, E. J., AND OTHERS. *Radiations from radioactive substances* 153
RYDZEWSKI, J. R., ED. *Introduction to structural problems in nuclear reactor engineering* 171

SABEL, C. S., COMP. *UKAEA and associated British work in controlled thermonuclear reactions* 217
SACHS, R. G. *Nuclear theory* 152
SACKS, J. *Isotopic tracers in biochemistry and physiology* 194
SADDINGTON, K., AND TEMPLETON, W. L. *Disposal of radioactive waste* 175
SANDERS, J. H. *Fundamental atomic constants* 156
SAXE, R. F. *Approach to thermonuclear power* 216
SCHUBERT, J., AND LAPP, R. E. *Radiation* 197
SCHULTZ, M. A. *Control of nuclear reactors and power plants* 172
SCIENTIFIC EQUIPMENT CO. *Isotope index* 195
SEABORG, G. T. *The transuranium elements* 135
SEGRE, E., ED. *Experimental nuclear physics* 151
SEMAT, H. *Introduction to atomic and nuclear physics* 151
SHANKLAND, R. S. *Atomic and nuclear physics* 151
SHARPE, J. *Nuclear radiation detectors* 199
SHILLING, C. W., ED. *Atomic energy encyclopaedia in the life sciences* 196
SHUMILOVSKII, N. N., AND MEL'TTSEN, L. V. *Radioactive isotopes in instrumentation and control* 193
SIM, D. F. *The future of national and international regulations covering the use of ionizing radiations* 198
SIMON, A. *An introduction to thermonuclear research* 216
SINGLETON, W. R. *Nuclear radiation in food and agriculture* 194
SKEATS, N. B., AND MULLINS, P. R. *Particle accelerators: selected references to books and articles* 155
SLATER, D. N. *Gamma-ray radionuclides* 156
SMITH, J. ROLAND, ED. *Guide to UKEA documents* 30
SMITH, M. L., ED. *Electromagnetically enriched isotopes and mass spectrometry* 192
SMULLEN, W. C., ED. *Basic foundations of isotope technique* 195
SMYTH, H. D. *A general account of the development of methods of using atomic energy for military purposes under the auspices of the U.S. Government, 1940—45* 136
SNELL, A. H., ED. *Nuclear instruments and their uses* 199
SOCIETE FRANÇAISE DE METALLURGIE and COMMISSARIAT A L'ÉNERGIE ATOMIQUE. *Plutonium 1960* 174
SOLOMON, A. K. *Why smash atoms?* 155
SPEAR, F. G. *Radiations and living cells* 197
SPINKS, J. W. T., AND WOODS, R. J. *An introduction to radiation chemistry* 196
SPITZER, L. *Physics of fully ionized gases* 215

SPRING, K. H. *Photons and electrons* 154
STANG, L. G. *Hot laboratory equipment* 195
STARR C., AND DICKINSON, R. W. *Sodium graphite reactors* 135
STEPHENS, W. E., ED. *Nuclear fission and atomic energy* 153
STEPHENSON, R. *Introduction to nuclear engineering* 171
STICHTING VOOR FUNDAMENTAL ONDERZOEK DER MATERIE. *Jaarvesslagen* 92
STIX, T. H. *Theory of plasma waves* 215
STROMINGER. D., AND OTHERS. *Table of isotopes* 156
STUART, D. F. O., AND BIRCH, F. *Survey on the use of radioisotopes in British industry and industrial research* 193
STUDIENGESELLSCHAFT ZUR FÖRDERUNG DER KERNENERGIEVERWERTUNG IN SCHIFFBAU UND SCHIFFAHRT. *Jahrbuch* 91
SUBE, R. *Dictionary of nuclear physics and technology* 134
SUDERSHAN, E. C. G., AND OTHERS, EDS. *Proceedings of the 1960 Annual International Conference on High Energy Physics at Rochester* 154
SWALLOW, A. J. *Radiation chemistry of organic compounds* 196

TAIT, J. H. *Neutron transport theory* 154
TAUBE, M. *Plutonium* 174
TAYLOR, D.
  The measurement of radioisotopes 199
  Neutron irradiation and activation analysis 193
TAYLOR, J. M. *Semiconductor particle detectors* 199
THEWLIS, J., AND OTHERS, EDS. *Encyclopaedic dictionary of physics* 134
THIRRING, H. *Power production* 170
THOMPSON, W. B. *An introduction to plasma physics* 215
THOMSON, SIR G. *The atom* 135
THOMSON, J. J. *The conduction of electricity through gases* 154
THORNDIKE, A. M. *Mesons* 154
THRING, M. W., ED. *Nuclear propulsion* 173
TITTERTON, E. W. *Facing the atomic future* 135
TOBOCMAN, W. *Theory of direct nuclear reactions* 153
TOKYO UNIVERSITY. INSTITUTE OF NUCLEAR STUDIES. *Annual report* 91
TOLANSKY, S. *Introduction to atomic physics* 151

UMAN, M. A. *Introduction to plasma physics* 215
UNITED KINGDOM ATOMIC ENERGY AUTHORITY.
  Film catalogue 30
  Glossary of atomic terms 134
  Guide to UKAEA documents 30
  The nuclear energy industry of the U.K. 31
  The United Kingdom Atomic Energy Authority: its history and organisation 31
UNITED KINGDOM ATOMIC ENERGY AUTHORITY. CULHAM LABORATORY.
  An annotated bibliography of articles on plasma physics and controlled thermonuclear research by UKAEA staff 213
  Culham Laboratory for plasma physics and fusion research 213
  Reports, etc. issued up to June 1964 213
UNITED KINGDOM ATOMIC ENERGY AUTHORITY. RADIOCHEMICAL CENTRE.
  Catalogue of radioactive products 195
  Radioactive isotope dilution analysis 193
  Selected references to tracer techniques 192
  The radiochemical manual 195

UNITED NATIONS.
  Atomic energy: glossary of technical terms 134
  Report of the United Nations Scientific Committee on the Effects of Atomic Radiation 196
UNITED NATIONS ATOMIC ENERGY COMMISSION.
  An international bibliography on atomic energy 131
UNITED STATES ATOMIC ENERGY COMMISSION.
  A.E.C. news release index 44
  Atomic energy facts 43
  Atoms for peace series 135
  Availability of nuclear science conference literature 136
  Bibliographies of atomic energy literature issued or in progress 44
  Bibliographies of interest to the atomic energy program 44
  Civilian nuclear power 170
  Controlled fusion and plasma research 217
  Controlled thermonuclear processes 217
  Controlled thermonuclear reactions 217
  Effects of radiation and radioisotopes on the life processes 196
  Fossil fuels in the future 170
  Fundamental nuclear energy research 43
  Guide to abstracting and indexing for N.S.A. 44
  Hot laboratories 195
  Index to conferences relating to nuclear science 136
  Index to press releases 44
  Major activities in the atomic energy program 43
  Motion picture film library 43
  Nuclear reactors built, being built or planned in the U.S.A. 172
  Organisations having publications exchange agreements with the A.E.C. 43
  Power reactor development programs 170
  Proceedings of technical meetings 136
  Radioactive fall-out 197
  Radioactive waste processing and disposal 175
  Radioisotopes in agriculture 194
  Radioisotopes in medicine and human physiology 194
  Radioisotopes in science and industry 193
  Reactor safety 173
  Report number series used by the Division of Technical Information Extension 43
  Reprocessing of irradiated fission reactor fuel and breeding material 173
  Rules and regulations 198
  Selected readings on atomic energy 131
  Selected reference materials on atomic energy 135
  Selected technical translations 137
  Shipping, handling and storage of radioactive materials 198
  Subject scope of N.S.A. 43
  Subject headings used by the U.S.A.E.C 44
  Technical books and monographs 43, 124
  Technical information services of the U.S.A.E.C. 44
  Translation title list and cross reference guide 138
  Understanding the atom 43
  What's available in the atomic energy literature 44
UNITED STATES ATOMIC ENERGY COMMISSION. DIVISION OF ISOTOPES DEVELOPMENT.
  Radioisotopes in world industry 192
  Special sources of information on isotopes in industry, agriculture, medicine and research 192

UNITED STATES ATOMIC ENERGY COMMISSION. NAVAL REACTORS BRANCH. *The Shippingport pressurised water reactor* 135

UNITED STATES DEPARTMENT OF LABOUR. *Employment opportunities in the atomic energy field* 43

UNIVERSITY OF CALIFORNIA. LAWRENCE RADIATION LABORATORY.
Bibliography on controlled thermonuclear processes 214
Controlled thermonuclear research: semiannual report 213
Status report 214

UNIVERSITY OF SYDNEY. WILLS PLASMA PHYSICS DEPARTMENT. *Six-monthly progress report* 214

VAN NAME, F. W. *Modern physics* 151
VEALL, N., AND VETTER, H. *Radioisotope techniques in clinical research and diagnosis* 194
VITALES, B. *A bibliography on heavy mesons and hyperons* 154
VON ENGEL, A. *Ionized gases* 214

WALLACE, B., AND DOBZHANSKY, T. *Radiation, genes and men* 197
WALTON, G. N., ED. *Glove boxes and shielded cells for handling radioactive materials* 195
WARREN, F. H., AND OTHERS. *A growth survey of the atomic industry* 45
WASHTELL, C. C. H. *An introduction to radiation counters and detectors* 199
WATT, D. E., AND RAMSDEN, D. *High sensitivity counting techniques* 199
WAY, K., ED. *Nuclear data tables* 156
WEINBERG, A. M., AND WIGNER, E. P. *The physical theory of neutron chain reactors* 172
WEINSTEIN, R., AND OTHERS. *Nuclear engineering fundamentals* 171
WENDT, G. *Prospects of nuclear power and technology* 171
WHITE, G. N. *Principles of radiation dosimetry* 136
WHITEHOUSE, W. J., AND PUTMAN, J. L. *Radioactive isotopes* 192
WILETS, L. *Theories of nuclear fission* 153
WILHELM, H. A., ED. *The metal thorium* 174
WILKINSON, D. H. *Ionization chambers and counters* 198
WILKINSON, W. D. *Uranium metallurgy* 174
WILKINSON, W. D., ED. *Extractive and physical metallurgy of plutonium and its alloys* 174
WILLIAMS, I. R. AND M. W. *Basic nuclear physics* 152
WILLIAMS, K., AND OTHERS, EDS. *Radiation and health* 197
WILLIAMS, W. S. C. *An introduction to elementary particles* 154
WILSON, R. R., AND LITTOVER, R. *Accelerators* 155
WOLFENDALE, A. W. *Cosmic rays* 155
WORDSWORTH, A. D. *Nuclear fuel handling* 174
WRAY, E. T., COMP. *Radiation protection: reference list of official and semi-official publications* 198
WU, T., AND OHMURA, T. *Quantum theory of scattering* 153

YAGODA, H. *Radiation measurements with nuclear emulsions* 199
YANG, C. N. *Elementary particles* 155
YARWOOD, J. *Atomic physics* 151
YATES, B., COMP. *Bibliography on nuclear propulsion for ships* 173
YIFTAH, S., AND OTHERS. *Fast reactor cross sections* 172

INDEX TO PERIODICALS

Abstracts of selected articles from Soviet Bloc and Mainland China technical journals 53
ACEC review 92
Achievements in science, Moscow 48
Acta isotopica 199
Acta radiologica 200
Advances in nuclear science and technology 175
American journal of roentgenology 200
American Nuclear Society transactions 176
Analytical abstracts 20
Annals of physics 156
Annual review of nuclear science 138
Applied atomics 138
ASEA journal 92
ASEA research 92
Atom, London 30
Atomdokumentation, Nyköping 91
Atomic energy clearing house 138
Atomic energy in Australia 32
Atomic energy law journal 138
Atomic energy review 112
Atomics 138
Atomkernenergie 138
Atomkernenergie Dokumentation 49, 127, 132
Atomki Közlemények, Debrecen 52, 54
Atomnaya energiya 47, 137
Atomo e industria 90
Atompraxis 138
Atoms in Japan 91
Atomtechnikai tájékoztató, Budapest 54
Atomwirtschaft 138
Atoomenergie 92

Beiträge aus der Plasmaphysik 218
Belgicatom bulletin d'information 92
Bibliographie scientifique hebdomadaire, Saclay 59
Bibliography of geological literature on atomic energy raw materials 176
Bilten dokumentacije, Belgrade 54
British journal of applied physics 156

British journal of radiology 200
Bulletin d'information A.T.E.N. 89
Bulletin d'information sur les applications industrielles des radioelements 200
Bulletin d'information technique, Brussels 78
Bulletin d'informations scientifiques et techniques 89
Bulletin de l'Association Suisse pour l'Énergie Atomique 93
Bulletin of the American Physical Society 156
Bulletin of the Atomic Scientists 138
Bulletin of the U.S.S.R. Academy of Sciences 137
Bulletin signalétique 132

Calandrier international des conférences, congrès, etc. 136
Canadian journal of physics 26, 156
Canadian nuclear technology 139
G.E.G.B. digest 31
CERN courier 114
Chemical abstracts 20, 131
Comptes rendus hebdomadaires des séances de l'Académie des Sciences, Paris 156
Contemporary physics 157
Courrier de l'A.T.E.N. 89
Czechoslovak journal of physics 54

Doklady Akademii Nauk S.S.S.R. 136, 137

Energia es atomtechnika 52, 139
Energia nuclear, Madrid 139
Energia nucleare, Milan 64, 139
Énergie nucléaire, Paris 139
Engineering index 132
Euratom bulletin 113
Euratom information 113
Euronuclear 139
Europa nucleare 139
European community 113
Express information, Moscow 48

Fizika tverdogo tela 136
Foratom informations 114
Forthcoming international scientific and technical conferences 136
Forum memo 45

General Electric Co. Research Laboratory Bulletin 214
Gen-ken, Tokyo 90

Health physics 200

Industries atomiques 139
Institute of Electrical and Electronic Engineers, Professional Technical Group on Nuclear Science. Transactions 139
Institute of Nuclear Sciences "Boris Kidrich". Bulletin 54
Instrument construction 137
International Atomic Energy Agency.
　Bulletin 112
　Conferences, meetings, training courses 136
　Library accessions list 112
　List of bibliographies on nuclear energy 112
　List of references on nuclear energy 112
International journal of applied radiation and isotopes 200
International journal of radiation biology 200
Isotopes and radiation technology 38
Izvestiya Akademii Nauk SSSR 137

Jaderna energie 54
Joint Research and Development Program quarterly digest 113
Journal de radiologie 200
Journal of analytical chemistry of the U.S.S.R. 137
Journal of applied chemistry 20
Journal of applied chemistry of the U.S.S.R. 137
Journal of fluid mechanics 218
Journal of inorganic and nuclear chemistry 200
Journal of nuclear energy, parts A and B 176

Journal of nuclear energy, part C 218
Journal of nuclear materials 176
Journal of nuclear science and technology 139
Journal of quantitative spectroscopy and radiative transfer 218
Journal of radiation research 200
Journal of research of the National Bureau of Standards 41
Journal of scientific and industrial research, New Delhi 32
Journal of scientific instruments 139
Journal of the Atomic Energy Society of Japan 91
Journal of the British Nuclear Energy Society 176
Journal of the Joint Panel on Nuclear Marine Propulsion 176
Journal of the Physical Society of Japan 218

Kakuyugo Kenkyu 214
Kernenergie 140
Kristallografiya 136

Library of Congress. Monthly index of Russian accessions 53

Magyar Tudományos Akadémia Központi fizikai 54
Metallurgical abstracts 20
Minerva nucleare 201
Monthly catalog of U.S. Government publications 40

National Bureau of Standards. Technical news bulletin 41
National Lending Library. List of books received from the U.S.S.R. 54
National Lending Library. Translations bulletin 137
Neue Technik 93
New scientist 140
Notes d'informations, Paris 89
Notiziario del CNEN 90
Nuclear energy 140
Nuclear engineering 176
Nuclear engineering abstracts 131
Nuclear fusion 218
Nuclear India 32

Nuclear industry 45
Nuclear instruments and methods 140
Nuclear medicine 201
Nuclear news 140
Nuclear physics 157
Nuclear power patents bulletin 176
Nuclear safety 38
Nuclear science abstracts 20, 26, 39, 49, 127, 131
Nuclear science abstracts from Japan 90
Nuclear science and engineering 176
Nucleonics 140
Nucleus, Sydney 32
Nukleonik, Berlin 140
Nukleonika, Warsaw 54
Nuovo cimento 157

Physical review 157
Physical review letters 157
Physics in medicine and biology 201
Physics letters 157
Physics of fluids 218
Physindex 89
Planetary and space science 218
Plasma physics, accelerators, thermonuclear research 218
Power reactor technology 38
Priborostroenie 137
Proceedings of the Physical Society 157
Proceedings of the Royal Society 157
Progress in elementary particle and cosmic ray physics 157
Progress in nuclear physics 157
Progress of theoretical physics 157
Propriété industrielle nucléaire 89

Radiation botany 201
Radiation research 201
Radiochimica acta 201
Radioisotopes 201
Radiokhimiya 137
Radiological health data 201
Radiology 201
Reactor fuel processing 38
Reactor materials 38
Reactor science and technology 176
Reaktorn 91
Referativnyi zhurnal 53, 119
Reports on progress in physics 157
Research reactor journal 176
Review of scientific instruments 140
Reviews of modern physics 158
Rivista di ingegneria nucleare 90
Russian technical literature 138

Science abstracts 20, 131
Science, East to West 138
Scientific and technical aerospace reports 44
Scientific and technical news service (U.K.A.E.A) 30
SLA list of translations 137
Soviet abstracts 49
Soviet journal of atomic energy 137
Soviet national bibliography 50
Soviet physics—crystallography 136
Soviet physics—Doklady 136
Soviet physics—JETP 136
Soviet physics—solid state 136
Soviet physics—technical physics 136
Soviet physics—Uspekhi 136
Strahlentherapie 202

Technical translations 137
Teknisk ukeblad, Oslo 93
TNO nieuws, The Hague 92
Transatom bulletin 103, 137
Translations monthly 137

U.K.A.E.A. list of publications available to the public 30
U.S. Government research reports 41, 131
U.S. Naval Research Laboratory. Report of NRL progress 214
Uspekhi fizicheskikh nauk 136

Vacuum 140
Le Vide 140

World list of future international meetings 136

Zeitschrift für Naturforschung 72, 158
Zeitschrift für Physik 158
Zentralblatt für Kernforschung und Kerntechnik 132
Zhurnal analiticheskoi khimii 137
Zhurnal eksperimental'noi i teoreticheskoi fiziki 136
Zhurnal fizicheskoi khimii 137
Zhurnal neorganicheskoi khimii 137
Zhurnal prikladnoi khimii 137
Zhurnal tekhnicheskoi fiziki 136, 218

Made in Great Britain

Z
5160
A46

DEC 14 1966

RAYMOND H. FOGLER LIBRARY
DATE DUE

BOOKS ARE SUBJECT TO
RECALL AFTER TWO WEEKS